THE PEBBLE *in the* Shoe

Detecting the Causes of Distress and Pain in the Human Body

THE PEBBLE *in the* Shoe

Detecting the Causes of Distress and Pain in the Human Body

JOHN P. STREICHER with Karen B. Alexander

WINEPRESS WP PUBLISHING

Streicher Publications
P.O. Box 765
Flagstaff, AZ. 86002

Printed in the United States of America

Packaged by WinePress Publishing, PO Box 428, Emumclaw, WA 98022. The views expressed or implied in this work do not necessarily reflect those of WinePress Publishing. Ultimate design, content, and editorial accuracy of this work are the responsibilities of the author.

ISBN 1-57921-266-2
Library of Congress Catalog Card Number: 99-68539

DEDICATION

The rights of children are distressingly fragile. They are dependent upon adults to a frightening degree. In the modern world the law attempts to define and protect the rights of children. It attempts to assure them of a minimum of care and to prevent more gross forms of abuse. Yet there are many subtle ways in which the rights of a child can be invaded.

This book is dedicated to protecting the rights of children and adults.

To a friend Betty who helped with our work.

John

P.S. Her husband helped proofing.

6/2000

Acknowledgments

Special thanks go to the family and friends who have made this book possible through their encouragement and expertise. There are many not listed here who have been supportive, and there are countless patients and students whom I wish I could mention. I hope it is sufficient to say thank you to all of you.

Hattie Szczpansky (Aunt Hattie)
Rosalie Streicher
Thomas W. Alexander
Margaret Flood
Arthur Berke, D.D.S.
Andrew Loomis, D.D.S.
Donald Yoshikawa, D.D.S.
Jay Vorah, D.D.S.
Ernestine Dunlap, speech pathologist
Eloise Shields, school psychologist
Louis Kaplan, Ph.D., school administrator
Ken Flood, school administrator
James E. Barlow, science writer/editor
Kenneth Stratford, medical graphic artist

Contents

Part Two: The Pebble in the Shoe (Finding Causes)

Part Three: Therapy

Part Four: Some History

Preface

I liken John Streicher to Sherlock Holmes, Jonas Salk, and the great men of the Bible.

John has the gift of a scientific mind, which has given him the capability of methodically solving the *what*-dun-it cases that have baffled professionals in education, medicine, dentistry, and the rehabilitation therapies. Using logical thinking and the powers of observation that Sir Arthur Conan Doyle attributed to his famous character Sherlock Holmes, Streicher has detected what causes most speech and learning problems, detrimental dental conditions, crooked form, jaw dysfunction, tension-pain patterns, and numerous other symptoms.

The culprit turns out to be nobody but the sufferer himself. Without realizing it, patients with medical and dental symptoms and students with speech and learning disorders are snared in habits that cause their bodily systems to alter their function and growth patterns to survive.

Sleuth Streicher looks at the body's form and observes the way it functions for speech as crucial clues in detecting what habits lurk behind the struggles. I often felt like a Dr. Watson as I initially missed these clues and made illogical assumptions. I hold a master's degree in one of the fields based upon assumptions and equipped only for symptom treating. I am a speech pathologist heavily indoctrinated into teaching ways for people to compensate for their speech and learning difficulties. As one of Streicher's first students, I was called to the carpet repeatedly for trying to fill "old wineskins with new wine." To become cause-oriented now that Streicher has uncovered causes requires a new way of thinking.

I liken Streicher to Jonas Salk, who discovered the polio vaccine, because Streicher found an equally unexpected form of cure. He found a vaccine that serves not only to prevent but to cure habit-caused disorders, and one that is readily available to all mankind without cost.

I liken Streicher to the faithful men of the Bible, because he asked the Creator of mankind for guidance in his discoveries, and he considered God's design to determine how we are suppose to talk, chew, swallow, and move our jaws. Streicher has suffered four decades of persecution for his diligent pursuit and success. Professionals who have feared losing their occupations or changing their ways have fought Streicher's discoveries, and they have discouraged his followers, who must work in secret or likewise be persecuted.

Streicher's discoveries will require a major alteration in our medical and educational approach to symptoms. If sufferers demand to know the cause of their problems and to be provided permanent relief, symptom management and remedial methods will become old models of treatment.

In this book Streicher tells how his discoveries unfolded; he presents a new paradigm—a new set of principles for finding causes and curing many conditions that trouble us, and he shares numerous case studies.

Streicher illustrates his detection methods with a metaphor about an undetected pebble in a shoe. In chapter one, Sleuth Streicher tells how it all began—with simple observations.

Karen B. Alexander

Part One

Introduction of Principles and Cases

Chapter One

Aunt Hattie's Introduction

My aunt taught me to make the letters of the alphabet with my fingers when I was a little boy back in the 1930s. I heard people call her "deaf and dumb." I hated that. I knew Aunt Hattie was not dumb.

But I think she was a true paranoid. She didn't imagine people talking about her and laughing at her behind her back. I witnessed them doing it. I hated that too.

No one ever taught my aunt to talk. I could not understand why she could not learn. She had started out talking until a severe infection at age three destroyed her hearing. From what I could see, she had the equipment for speech like anyone else. Why couldn't she talk?

I met the caretaker for the apartments where I lived. I struck up quite an acquaintance with him. He wrote me notes, and he used some sign language. I noticed he acted much like my aunt. Both moved their mouths as they signed.

My aunt and the caretaker seemed to set my destiny to prove that deaf people can be taught to talk—clearly and with near normal voice quality. I became a speech pathologist. When anyone asks why, I start by telling them about Aunt Hattie and others—how they influenced my decision to choose speech pathology for a career. It wasn't a likely career for me. I am most interested in and gifted in science. It was Aunt Hattie's silent screams that summoned me to turn speech pathology into a scientific field. I wanted to find the cause and cure of speech disorders.

I have done it.

In this book I tell *how* I did it. I share my observations, research, experiences, and treatment program. I tell how discovering the cause and the cure of speech disorders led to discovering the cause and cure for many dental and jaw problems, tension and pain patterns, and hearing and learning disorders. Let me tell you how the discoveries unfolded.

It began five decades ago. I was in China at the end of World War II. I noticed the behavior of deaf-mute people there. They exhibited paranoid behavior like Aunt Hattie, and rightfully so. People made fun of them. It seemed their lot in life was no better than my aunt's. I could understand why the deaf needed to migrate to each other for compassion and understanding.

As I awaited my discharge from the Marine Corps in 1947, I visited a friend of mine who had become a speech pathologist. I went to some of her classes to see how the therapy worked. I witnessed something that made my heart sink.

One of her speech students, a little boy dressed in a sailor's outfit, was as cute as a bug's ear. I watched as he tried and tried and tried to do what my friend asked of him. He tried until the tears came. Then he just sat there as his therapist turned her attention to another student. After the children left the room, my friend said, "The one in the sailor outfit just didn't try hard enough." That nearly broke my heart. I can tell when someone is trying and when someone is being lazy. I'd worked with enough goof-offs in the military. That little boy was no goof-off. The speech therapy method was goofy. I didn't like it.

I considered other fields of study at this time and signed up for a few college classes only to be called back into service for the Korean War. In Korea I had the chance to observe people under much stress and strain. I particularly noticed their oral habits—the many things they did with their mouths. I stored these observations in the back of my mind.

When I returned home, I signed up for mostly science classes at a junior college. After that I decided to enroll in speech pathology at the University of Southern California. (I tell about my educational experiences in chapter ten, The Speech Therapy Field.)

I assumed my first speech position after college with the school district of Torrance, California, in 1959. During my first year I tried the speech methods that I'd been taught. Some children improved, but they weren't necessarily the children who worked hard. In fact, some who goofed off improved the most. Other children did not improve significantly but were trying very hard. This distressed me.

"I want to quit," I told Dr. Kaplan, the superintendent.

"Why? You have one of the best records of anybody here. Why would you want to leave?" he asked.

"I don't know what I'm doing. There is no way to predict through this type of therapy which children will be able to correct their speech and which ones won't. I have kids climbing the drapes, doing everything but what I ask them to do, and their speech is improving. The kids who do what I ask and who I think are doing a great job are not improving. We have no way of knowing with this therapy. Yet our results look good on paper because some kids are improving."

I further explained that a study needed to be done to figure out what causes the speech problems we see in children. Dr. Kaplan liked the idea. He said, "We have some money in the school district that could be used for a study. Would you set it up? You will have to work with the speech supervisor, Ernestine Dunlap."

I agreed. Ernestine showed interest in my theory. I was becoming more and more convinced that oral habits cause speech disorders. She had been looking into it with me. She too was not happy with speech therapy methods. We had decided that something should be done about it. We reviewed speech therapy research and discovered that most of it was a farce. Many findings had been falsified. Most of the research that professors had done in order to get their doctorates could not be verified. The speech field was awash in false information.

We started our research with four kindergarten classes, each with about thirty children. We had an influx of multiculturalism. There were Chinese, Japanese, Indonesian, Micronesian, Melanesian, Black, and Caucasian students.

Eloise Shields, the school psychologist, was delighted with the proposed study and suggested that we first study the children to see if there was any psychological reason for the speech problems. We took the study into several schools, testing about 240 children. We planned to study them for three years (1962–1965). But by the end of the second year, the number of children who moved in and out of the district disrupted the study.

We observed tremendous physical damage of the oral structure (mouth and surrounding areas) of these children. We also saw oral habit patterns that could be associated with the structural damage. To demonstrate this I asked the school dentist, Dr. Chalis, to step into my therapy room to see six children whom I had gathered from several classrooms. Two of the children were Black, one was Chinese, one was Japanese, one was Spanish, and one was Caucasian. When the dentist entered the door, I asked the children to stand up and say hello to him. A big grin came over the dentist's face. He walked out. I could hear him laughing outside the door.

I dismissed the children, and Dr. Chalis came back in. He said, "I wouldn't have believed it. They all looked alike."

"Do you know what their habit is?" I asked.

"No," he said.

I explained that each of the children—all of whom looked somewhat like chimpanzees because of their widespread mouths—sucked on four fingers.

He said, "But they were white, black, and yellow—and looked like brothers and sisters. I would never have noticed this, because in a dental practice I see one person at a time. You are fortunate to get to see children in large numbers."

Later on Dr. Chalis would be very helpful as we tried to come up with devices to help reeducate the tongue for correct function. This was the next step after discovering that oral habits did indeed cause the varying speech patterns and varying dental structures we found in children.

We made all sorts of dental devices, things that we thought might recondition the tongue. For one child we made a sharp metal plate and placed it between his lower teeth and tongue. His tongue learned to put a little dent into itself to avoid the sharp metal. After four days we removed it to see how long this conditioning would last. It did not last long enough to snap a picture. It would be a while longer before I would find the answer.

Meanwhile teachers became interested in our work, and in 1965 I began experimenting with kindergarten children and phonics. I got some opposition from people who believed that children this age could not learn phonics. But teaching them to read by phonics turned out to be very successful. We taught kindergarten children word attack skills and had them reading sentences in no time. Today that approach is well formulated and used at the kindergarten level in many schools.

One reason for working with the kindergarten children was to get them comfortable with me. When children are comfortable, they will expose their oral habits. I spent much time observing their behaviors. But one little girl baffled me.

This Japanese girl would not say a word to me. I could not get her to talk at all. Then one day she was standing in the hall with her peers. She looked up at me and said something that I could not understand. I had to ask the other children what she wanted. They knew. "She wants to know if she can get a drink of water."

I told the child to get herself a drink. Then I slipped into the classroom and asked the teacher, Mrs. Harper, about the child's speech. Mrs. Harper could not understand anything the child said either. She said, "But her classmates understand everything. They have cracked her code."

I thought about the research that has been done on twins. Twins are able to develop their own language, and no one else knows what they are saying. But what had caused this little girl not to speak clearly in Japanese or English? She didn't demonstrate any oral habits in the classroom. But I saw a clue—other than her speech—to suggest that she habitually placed something in her mouth. She had green teeth. No one had been able to determine what was making them green. Her dentist had even referred her to the department of dentistry at the University of California. No one there could figure it out.

Christmas vacation came and went before I discovered the child's habit. When school resumed in January (1966), Mrs. Harper came looking for me right away. I was in the teacher's lounge where I gathered information about children. She came in and asked me to come to her room. There I witnessed the Japanese girl chatting with her friends. I could understand her speech. Mrs. Harper asked what I thought. I did not have the foggiest idea what happened.

I told Mrs. Harper I would return at my regular time to investigate this further. I decided to have a sharing time with the class. The children could take turns telling about their Christmas vacation. The little Japanese girl waved her hand excitedly and nearly jumped off the rug to speak, so I invited her to come up in front of the class first.

"My grandma came to visit us, and she sewed this yuck thing around my blanket."

I asked the little girl if her blanket was green. She looked at me with her eyes opened wide and said, "You've been in my bedroom!" I assured her that I hadn't, that I was just guessing.

The little girl's mother confirmed that Grandma took the child's shabby blanket and sewed a new acetate trim on it. The little girl didn't like it and spit it out. She didn't like the taste. Yes, it was a green blanket. I told the mother that I was pretty sure the blanket was responsible for staining the teeth.

Getting the habit stopped made a tremendous difference in the little girl's speech. But she needed to have her tongue reconditioned for a correct pattern. With the habit stopped, reconditioning was possible.

I knew by then that an oral habit causes problems by forcing the body to operate around it. Simply eliminating a habit would not bring the body back into balance. Something was needed to recondition the system. So I was looking for that something. And I was trying many different objects with no success. I resorted to prayer.

As I was driving down Torrance's Crenshaw Boulevard on my way to a school one morning, I talked to God about needing his help on this. Then in my mind I saw a curtain lifting. Before me stood a stage with words upon a screen, but I could not make them out. They were fuzzy—like when I take my glasses off and try to read. Then it cleared up enough for me to see one long word—*vaccination*.

Vaccination—when you use something akin to the body to protect or restore the body—like calling on the body's own immune system with just enough of a virus to ward off the illness that the virus can cause. I understood the idea, but what was I to find?

About 3:00 P.M. the same day, I sat in my therapy room with several children. Willie sat to my left, watching the clock and sucking his thumb, while I talked with a little girl about a picture she was drawing. Wanting to get her to talk freely, I said something about her nice picture and nudged Willie. I said, "Isn't that right, Willie?"

"Yes," Willie said with a perfect *s*-sound.

Willie had never made an *s*-sound correctly. He had a lateral lisp, which means he made sounds produced with air out the side of his mouth instead of the front. This makes the speech sound slushy like when cartoon character Daffy Duck spits out "suffering succotash."

I was stunned. I asked him to repeat. He took his thumb out and said "yes" like Daffy would. I asked him to put his thumb back in. "Yes" came out correctly again.

I began working with the thumb to see if it could indeed be the part of the body, the vaccination, I was to find. It made sense. The body would not reject a part of itself. I tried "thumb therapy" on several children who had not corrected their lateral lisps after several years of therapy. They corrected. Correcting lateral lisps became not only possible but easy.

I began to use the thumb to treat all kinds of speech patterns. It worked.

By this time I was getting quite a reputation in the school district for what I was discovering. Principals of schools where I did not work started calling, asking me to assist with special cases.

One such instance involved a girl being earmarked for an EMR (educable mentally retarded) class. The parents fought this placement. The principal wanted me to get them off his back.

My investigation of this case led me to understand how cloth over wire wears the edges of teeth. This child had a niche in one tooth, leaving the outer edges of it sharp and pointed. Her speech was bad; she talked like she had a hot potato in her mouth. I alerted the mother to be on the lookout for something the child kept in her mouth that held the tongue down.

About ten days later I received an envelope from the mother. I pulled out arched wires that looked much like the wires an orthodontist would use. The mother asked whether the wires could be the problem. She had pulled them out of the tattered ears of her daughter's stuffed bunny.

I went to see the little girl and placed the wires in her mouth. They fit according to her tooth wear pattern and the way she had conditioned her tongue to flatten. With the damaging habit stopped, I proceeded to teach the mother how to recondition the tongue with the thumb.

The child's speech corrected, and she started participating in the classroom. I checked on her two years later. She was handling the regular classroom work just fine.

But the school's speech pathologist and psychologist were not happy. They felt that I had made them look like they did not know what they were doing. They could not rejoice about the little girl's condition being treatable. They could not rejoice about the discoveries made in their own school district.

As the success of the therapy became more and more evident, the hostility from other professionals who did not want to change their practices became more than I wanted to endure. I did not feel like putting up with their whining. I could be my own boss. So in 1968, I bought a trailer and went around to the Catholic and Christian schools, setting up speech programs for them.

Margaret Flood, the mother of one student whom I helped, was very enthused about the therapy and wanted to be trained to do it. After I trained her, she began working with me. Until she retired, Margaret helped me provide therapy in schools and clinics, and she helped me conduct seminars and training programs to teach my diagnostic and treatment methods to others.

I also met Dr. Arthur Berke, an orthodontist who treated many of the children in the private schools. He approached me one day and asked, "What are you doing to my patients?"

I thought, *Boy, am I in hot water again.* I asked, "What do you mean?"

Dr. Berke said, "I have been seeing corrections in the children you are working with—children whom I have in bands and braces. I am getting positive changes that I haven't even planned for. What are you doing?"

I told him that I was finding their oral habits, getting them to stop the habits, and showing them how to incorporate a new pattern into their system for talking and eating. I was teaching them the right way to bite, the right way to chew food, and the right way to swallow in order to correct their speech and any related problems.

"And you have an office here?" Dr. Berke asked.

"Yes," I replied.

"Would you close it and move your practice in with me? You can use my office, you can use my equipment, and together we can compile the actual dental proof of what you are doing."

I thought about his offer. It would not interfere with my current work. So I agreed and joined him in his office in 1971. Through Dr. Berke I gained much knowledge of dentistry. He became my mentor through the discoveries that were to follow. We found many dental problems caused by oral habits. We made pictures and models, and we documented.

Throughout the book I talk about how we discovered that habits cause other problems—jaw problems, learning disorders, sleep-tension patterns, and headaches, for example. These conditions are related. They have a common cause—the things we do to ourselves habitually. And one habit pattern can cause numerous symptoms.

I have treated thousands of patients over the past forty years and have seen how this therapy improves lives, especially to those who suffer from tension-pain patterns, which I address in the next chapter.

Many patients' stories are told in this book. Their successes bring me great joy. They give a special purpose to Aunt Hattie's life, a purpose for her suffering.

Chapter Two

Pain

The root of pain is not always easy to dig up.
—James E. Barlow

"I've always had headaches but nothing like what I have suffered the past few months," Meg told me. "I've been to a physician, a dentist, and a psychologist. They did not find a reason for my pain."

I asked Meg if anything different had happened in her life recently. Meg smiled and said, "Yes, I just got a promotion at work. I am now in charge of my section.

"I really am thrilled about this promotion, but it is so much more stressful than my other job. I need to correct people's work, you know, point out their mistakes, but it bothers me to do this. So on top of doing my own work, I take their work and make the corrections myself."

Meg thought taking on the extra work would be less stressful than confronting her staff. But then she found herself avoiding all situations that involved conflict. "I've been avoiding dates lately, because I can't even enjoy them," she said. "I'm desperate now. I don't want to lose my job or give it up. I'm taking prescription pain medicine, aspirin every four hours at work, and at least two stiff drinks every night when I get home. Can you help me?"

I told Meg that I could. I saw that she tightened shoulder, neck, and head muscles in response to stress. It was a learned pattern of uniting physical tension with emotional stress. Until now she had kept the emotional stress at a level that gave her only temporary muscle pain that she could tolerate.

When her stress increased, however, it put her into a pain pattern that was almost self-sustaining. That is, she entered a cycle of pain that produced more pain. It took at least two to three days for one cycle to run its course.

First Meg had to learn to recognize what muscles she was tightening to meet her stress. Then she needed to learn to divorce her emotional stress from her physical response.

Meg, like many other people, had been told that she had to learn to relax, but no one could show her how. She had also been told to get rid of the stress in her life, but this is not a reasonable expectation for Meg or anyone else. In fact, no one can escape the daily stresses of life, because stress includes joy as well as sorrow and excitement as well as anger or fear.

Meg did not understand the difference between muscle tension and mental stress. She needed to learn that when her body responds to mental stress with physical tension, pain can result. In fact, most headaches and jaw and neck pains we suffer are brought on by muscle tension patterns. When excited or upset, we tense up physically. That is, we tighten muscles to protect us from emotional blows.

We do need to make a wall of muscle to protect our bodies from physical assault. But when we use that same wall of muscle to handle emotional situations, an unhealthy pattern evolves. This starts very early in a toddler's life and continues into adulthood.

Fortunate is the child who never learns this pattern. Unfortunate is the child who does not learn mental strategies to deal with mental problems; that child is the one who uses physical muscle tension instead. And this usually is learned by imitating family members.

We need to divorce mental stress from physical tension. The misuse of muscle tension results in muscles being held in a tight position for a long period of time. This causes interruption of blood flow to the muscles. Without blood flow, the muscles have no way to rid themselves of acid buildup. Acid buildup causes everything from severe to burning pain, even migraine-like headaches, which in turn can cause severe stomach disorders.

This was true for Meg. She developed stomach upsets with her migraine-like headaches, causing her to miss work and social events.

When people in pain, like Meg, let go of their physical tension patterns, they recall previous experiences that served to link emotions to muscle tightness. They also experience physical sensations as they start to relax the muscles. These include sensations such as hot, cold, tingling, twitching, and itching. There also may be spasms or muscle jerks like those that occur when falling asleep.

Therefore, to learn to relax is to focus on the muscles we have identified in our habit pattern. We need to concentrate on loosening them and permitting all the uncomfortable or painful sensations to emerge.

With the muscles relaxed, we then learn to keep them relaxed while doing routine tasks. For example, the person who kept the jaw tight habitually now focuses on keeping the jaw relaxed while pushing the vacuum cleaner, washing the dishes, taking a shower, shaving, and listening to others tell their problems.

Learning to relax is not without struggle. Our physical bodies have a defense mechanism. They fight to keep us in tension. They tend to want to cut off the sensations.

We must all know people who realized that they were tightening muscles and hurting themselves. They were able to self-correct and alleviate their pain. But many of us do not know what we are doing to ourselves and need to be taught to relax like Meg.

Many do not understand that relaxing does not mean going to sleep. To be a relaxed person is to be someone who can function—who can carry out daily tasks or special activities—without tension. To help Meg understand this, I presented the following scenario:

Two Accountants

Let's imagine a day in the lives of two very competent accountants who have similar problems at home and do similar tasks at work. Both go to work worrying about their situations at home, and both face a busy workday.

At the end of the workday, however, one goes home feeling exhausted and suffering aches and pains. This one cannot deal with the problems at home. To alleviate suffering, he relies on aspirin and liquor.

The other accountant is different. Unlike the partner who faces mental challenges with tension, this one never learned to add physical tension to mental problems. He goes home with enough stamina to tend to the mental challenges. There is—even at the end of the workday—enough time and energy to handle whatever he needs to do at home.

Those of us who never learned to physically tense up are blessed. Those of us who tense up have to unlearn our habit pattern.

After we let go of our tension pattern, one last uncomfortable feeling enters in. We may feel that something is missing. That something for Meg turned out to be a routine stop at a drugstore for pain relievers and a need every evening for two margaritas. But the painkillers and drinks were replaced with something better.

Meg got a further promotion to manager of her entire department. Once Meg learned not to tense up in response to her mental challenges, she could confront her staff members and provide the training they needed. In fact, Meg accepted a volunteer post as union representative that required public speaking.

The old Meg wouldn't have considered speaking in public. But as the new, relaxed Meg, she enjoyed the challenge.

Meg never considered returning to her old patterns. But one man I worked with did. I told Meg about him to emphasize the mental control a person can gain over physical tension. This other patient chose to remain in pain for a pending lawsuit. I'll call him Wayne.

Wayne's Pain for Gain

Wayne said he could not go into court without pain and convince the judge and jury that his employer had caused him unbearable pain. He knew he was not an actor. He needed the real pain for gain. Wayne actually had a good case because of his working conditions. His job required typing and talking on the phone at the same time while positioned in a chair that did not accommodate his size. He could not sit up straight. He had to hunch over the keyboard to perform his tasks. Supervisors ignored his repeated complaints.

Wayne had made it clear what was causing his discomfort. He had to work hunched over, looking down at a poorly angled monitor aimed for the convenience of a patrolling supervisor rather than for his comfort. Close working quarters prevented him from relaxing his arms and positioning his wrists correctly. These conditions caused wrist and finger problems as well as pain in the shoulders, neck, and head.

Wayne put up with the conditions for six months before filing a lawsuit. A dentist who knew about his physical pain referred him to me. I took Wayne through the same therapy program as Meg's, and it was as successful as Meg's in alleviating pain. Upon losing his pain, however, Wayne feared losing his court case.

He chose to go back to his old habits and reintroduce the pain until his court case was settled. He said, "I know what to do now, and I will do it just as soon as the case is settled. I do not want to live with this pain."

Wayne was happy with the results of learning to cure his pain. He won his case, took himself out of pain shortly after, and quit his job.

Both Meg and Wayne learned that mental activity is not pain producing because the brain has no pain receptors. They had suffered physical pain because of the way they handled mental stress. That is, they had been trained to meet emotional stress with physical tightness. They were cured of their pain by learning to give up control to gain control. When they let that tightness go, they gained mental control of mental stress. Both Meg and Wayne took charge of their lives.

In the next chapter, I introduce the cause of many other commonly suffered disorders and symptoms. I address how we start out in life—what allows some of us to develop problem-free and what causes many of us to struggle.

Chapter Three

Life Patterns

As the wind blows, so grows the tree.
—Streicher

What sets us on our life course? Why do some people seem to sail through life? They have good physical skills. They have no learning and speech problems. They have no major dental problems. They have few aches and pains. They are overall quite healthy, happy, and successful.

Others struggle with all these basic human functions. What sets them on such a bumpy course? Does the bumpy pattern start in the womb? Does it start in early childhood? Does it start later in life?

The bumpy pattern can start anytime. But it is worse when it starts in the womb and is left uncorrected. To understand why, let's look at what should happen in the womb before looking at what causes problems.

It is natural for the fetus to place a thumb in the mouth at a particular time. The growth and position of the fetus allow this to happen. The design of the body allows a little elbow to bend and a little thumb to reach the mouth. It is supposed to happen. There is a purpose.

The thumb works like a trigger to start bodily systems working together. It is the key for the life-sustaining systems (digestive, nervous, respiratory, circulatory, and endocrine) to work together.

Most babies seem to find their thumbs while in the womb, and, therefore, are born thumb suckers. Thumb sucking becomes the first source of pleasure.

Thumb sucking also prepares the child for breast-feeding and other human functions such as speech and language. It sets the course for natural development. The thumb sucker can begin using the oral structure (the mouth and jaw) to take in nourishment, to make sounds, and to begin learning from the environment upon birth.

Immediately after birth, physical signs tell us whether a baby sucked on a thumb or another part of its body. Suck marks on the skin show us if a baby missed or rejected the thumb and began sucking instead on a finger, foot, knee cap, arm, or any other body part that reached the mouth—including the umbilical cord.

When something other than a thumb sets up the swallow pattern and keys the bodily functions, the child starts life on a bumpy course.

The baby who veers from the intended path while still in the womb is the baby who encounters its first struggle upon birth. Its suck pattern is not set up for breast-feeding. The environment does not provide a nipple shaped like a fist or foot.

This baby needs to be introduced to its thumb. This will retrain its suck pattern. It will correct its swallow reflex and reintegrate body systems. Therefore I suggest that the current practice of using a pacifier, bottle nipple, or an adult's finger in any combination to stimulate a suck pattern be replaced with a practice of using the baby's thumb.

In some countries, mothers encourage babies to accept a pacifier. This too is unnatural and throws the body a curve.

The fetus that has used something other than a thumb as a triggering mechanism has a defective pattern established. The swallowing pattern is incorrect, and the system is not well integrated. The baby is not on the right course.

Can anything be done to put this baby on the right course? Yes, the best treatment is preventative—one that begins right after birth using the thumb.

Now let's turn to the individual who veers off the natural course sometime after birth. This is the person who acquires a habit.

By habit, I am talking about a habit that is done with the mouth—an oral habit. This is not to be compared to daily living habits or routines. Oral habits are consistent, persistent behaviors. There are no variations in their patterns. They are not random behaviors like fingers in the mouth one time and a teddy bear's ear another time. I am talking about habits that are done so persistently and consistently that the body is affected.

For example, a child who has an oral habit of sucking on a blanket causes a change in the way its body functions. While the blanket is in the mouth, the child continues to swallow, breathe, etc. But the way the child performs these basic body functions is altered. The body adapts to accommodate the habit.

Some habits also require the body to redirect its growth pattern. For instance, a child who either starts a four-finger suck habit in the womb or acquires it sometime after birth causes the dental arches to grow around the four fingers. The arches widen, and, consequently, the teeth come in widely spaced. The same habit can train the tongue to flatten, curl, twist, or pull back when the child swallows or attempts speech.

A look at the condition of the fingers, the mouth, and the way the jaw and tongue move during speech and eating gives me very obvious clues as to what the child does to cause its set of problems.

If the child acquired the habit after birth, correction is much easier than it is for one started in the womb. Generally an acquired habit undermines only part of the body systems. It does not undo all correct functions. Certain factors determine how much damage a habit does, however. These include: the shape and size of the object, how much of the oral structure is involved, and how much pressure is applied.

The child with the most serious problems is the one who got off course in the womb, then acquired a habit after birth. This child has compounding problems, but these problems can be cured.

I will describe many habits and the problems they cause in the following chapters. This is extensive information as oral habit patterns are as varied and numerous as the people on earth. There are similarities, but no two are exactly alike.

Take fingerprints as a comparison. They are similar, yet there is enough variation to identify an individual. Likewise, when it comes to habit diagnosing, no two individuals have identical habits or identical sets of problems. But there are similarities.

Having discovered that oral habits cause numerous problems with such variation, I do not see a need to use labels for describing conditions. To diagnose habit-caused ailments, the examiner can be very direct: "This is what you are doing to yourself, and these are the problems caused by your habits."

Therefore I will not use current medical, dental, psychological, or therapeutic terminology to describe symptoms caused by habits.

I will talk about individuals who suffer from impairments such as deafness, cerebral palsy, cleft palate, and retardation that are not caused by habits. Not only is it possible for these individuals to have found something in the womb other than the thumb, but many acquire habits after birth that compound their problems. They, more than anyone, need the edge of operating habit free.

Anyone operating around habits has to cope and compensate. This requires the use of intelligence or thinking ability to override the effect of the habit. That is, the child who has self-corrected or been taught to compensate for a deviant speech pattern uses intelligence to exert control over a motor skill. This causes an IQ (intelligence quotient) deficit that is not necessary when the motor skill is retrained for correct speech function. I will give an example.

A Japanese woman, whom I'll call Jane, came to me with her two young children who both had speech problems. They were incorrectly producing the *s, z, sh, ch,* and *j* sounds, which caused them to say things such as "Thee Thammy Thnake thwim" in place of "See Sammy Snake swim."

Using my therapy program, these children took only three months to correct their defective sounds. Shortly after they were dismissed from therapy, their mother visited me with concerns about herself.

Jane had noticed that her children could speak spontaneously; that is, without any effort. She explained that when she was in school, therapists helped her correct the same problem her children had, but her speech was not without effort.

Jane wanted to know why she had to think to say her words correctly. Her speech wasn't spontaneous like her children's; she had to control it. She said she could have no more than one drink at a party or she would lose the ability to control her speech.

Jane wanted to know what type of therapy I did that allowed her children to speak correctly without thinking. I explained the difference in my therapy and conventional speech therapy and suggested that her children could show her how to place her thumb in her mouth and practice. Jane seemed pleased, but as she was leaving she asked another question:

"Would this therapy work with singing?" She was taking a class in singing at a junior college and feared that she would fail because she could not project her voice.

I directed Jane to practice her speech sounds while singing and keeping her thumb in her mouth. Unfortunately her singing class finals were just one week away.

Jane did not have enough time to acquire the projection she needed to develop. With a thumb in her mouth, however, she perceived how much better her voice sounded. So Jane devised a little scheme.

When it came time for her to perform for the professor, she asked him to turn his back and just listen to her. He gladly obliged, and Jane proceeded to put her thumb in her mouth while she sang her selection. Later she informed me that she had passed her finals, was still practicing with her thumb, and was getting to where she could project well without it.

Jane found out that using the thumb in the mouth causes the air passageways to open up, the mouth to open up, and the tongue to go into the correct position, thereby allowing the voice to project full force from the diaphragm.

Jane is one example of how my therapy can help most voice patients as well as those who want to improve their singing or speaking ability.

Another example is a disc jockey whom I'll call Bert. We met at a dental office where I provided therapy. Bert had caps put on his front teeth because he was interested in becoming an actor and singer. He asked me why a speech pathologist worked in a dental office. Our conversation resulted in an appointment.

Bert wanted to increase his volume for singing, and he wanted help to overcome his reoccurring laryngitis. After the laryngitis condition corrected, we were able to begin work on his career concerns.

Bert displayed inappropriate facial mannerisms when he talked. He knew this would hurt his career. He practiced speaking with his thumb in front of a mirror to catch and stop these mannerisms.

As Bert improved his speaking appearance, he noticed he had some of the mannerisms when he sang. So he carried out the therapy when singing. This led him to notice that the therapy enhanced his voice quality. Bert was so pleased with his changes that he asked, "Would it hurt me if I practice with my thumb as a warm-up for speaking and singing? It helps me to relax and perform better at work. I could do it while driving, couldn't I?"

I said, "No problem. Just don't run into anybody!"

But he didn't listen to the last part. He "ran into" two girls standing on a corner as he waited for a traffic light to change. They giggled at him. Bert just turned to them and said, "It's okay. Someday you all will be doing this."

Bert saw how this therapy could help many people. His habits had kept him from doing what he wanted to do in life. It had been painful for him not to move ahead in his chosen career. He realized that many others suffered disappointments and failures from simple behaviors that they too were unaware of. Bert didn't hesitate to show off what had set him freely sailing on his true course in life. He had acquired, as had Jane, the freedom to speak, to sing, and to act without the extra baggage from oral habits.

Chapter Four

Learning

The road to learning is clearly marked.
It is the side trips that throw us a loop.
—Streicher

Learning starts before birth. We either start out learning on the course that nature provides in the womb, or we adopt another course.

To start out on the right course only requires that we sucked a thumb properly in the womb and did not acquire an oral habit before or after birth to throw us off.

Already at birth, we have selected the natural, intended neural pathway for learning or an unnatural neural path. The intended path sets us up for life's first successful event: feeding properly without stress. Infants whose systems were not set up with the proper thumb exercise will start life with frustration.

It is important to set these infants on the right course with proper thumb therapy right after birth—before they are introduced to feeding. This corrects the divergence from the natural path that was triggered in the womb. Thumb therapy steers these infants away from learning difficulties and reintegrates their systems for meeting their intellectual potential.

This does not mean that the child who sucked a thumb in the womb and the child who was retrained to do so after birth are on equal footing. There is a far greater tendency for the retrained child to go off course. Obstacles and difficulties could trigger a relapse to the pathway originally constructed. The child could go back to the habit established in the womb. This child needs to be watched to assure that the right road is maintained.

The child who had the thumb in the mouth while in the womb, likewise, has a tendency to return to the thumb for comfort. This is good. This should be allowed. This should not be replaced by pacifiers or anything else, which brings us to another point.

Babies should not be allowed to feed themselves with their bottles regardless of how timesaving this is for the caregiver. This can cause "self-feeding bottle syndrome," that is, mild-to-severe problems with learning, speech, and dental development as well as the rotting of teeth. (Baby bottles and pacifiers will keep remedial teachers, dentists, and speech pathologists in new cars for the rest of their lives!)

The child born competent for feeding—the one who did everything right in the womb—has established neurological patterns for being gratified through the feeding experience with the least amount of adjustment. This child's neurological learning development continues on a normal path that will benefit further learning.

The child born incompetent for feeding will begin the feeding process with trauma. To have a malfunctioning system retrained by a finger, pacifier, or bottle nipple starts setting up neuropathways for failure. Because this child is not introduced to its thumb for correct training, it returns to the womb habit to readjust its familiar neurological beginnings, where it met gratification with the womb object. This is the child of any race or language who will take neurological, speech, and learning difficulties associated with stress to the classroom.

These problems are universal. The same problems can be found in every culture, every race, every language. If a child has problems making certain sounds in one language, the same problems will occur in all languages spoken by that child. Likewise, when speech is cured for one language, it is cured for all other languages used.

The Kindergarten Experience

The child who never got off track will face school as a challenge and will be rewarded by his system in a way that he will achieve gratification from competition. This child and the struggling child can be picked out fairly easily in a kindergarten screening—a simple question-and-answer verbal screening taking no more than two or three minutes.

I have done this repeatedly with an accuracy of 90 percent. The ten percent I miss in screenings are those children with high IQ. They are able to cover up and compensate for their difficulties.

The child on the right course does not have to compensate. Everything works well. The bodily systems work in harmony. Potential can be met. A high intellectual capacity isn't needed for coping. There is enough intelligence to learn the necessary skills for reading, spelling, and language.

An IQ at the kindergarten level does not have a major effect in distinguishing capabilities. My research agrees with earlier findings, however, that boys do not develop as fast as girls do. Boys would have fewer problems if they started school a year later than girls.

Tackling First Grade

The foundation for skills crucial for future learning is laid in first grade. Children learn to listen and to pick up details as teachers read stories to them. Eyes and hands work at tasks to develop needed skills. Simple math concepts are introduced.

At this time struggling children realize that their coping skills are not functioning as well as most of their classmates. They may have already heard: "You talk funny," or "How come you are still sucking your fingers?" or "Why do you chew on your shirt?" This is when these children are confronted about their problems. They start to pull back. They no longer ask questions or respond in the classroom. They clam up. Or they become the completely obnoxious kids—the classroom problems. Their kindergarten teachers hardly recognize them as the children they had in class.

Rougher, Tougher Second Grade

Things get even tougher for the struggling children in second grade. Their speech problems already have created encoding and decoding problems for reading and language development. This became evident during first-grade academic activities.

These children still have a great time on the playground. But when the recess bell rings, they line up and are marched back to the dungeon. They know emotional stress, but they don't know why they can't perform in the classroom. They begin to hate it there.

School is a breeze for the children who are equipped to meet the challenges. For the children not equipped to meet the new onslaught, it is an uphill struggle all the way. They have to work harder and harder to stay in the game. But it is not too late.

Therapy Changes Learning Course

I developed a therapy program to put struggling individuals on the right course for learning and speaking. The therapy takes an individual back to where troubles began and helps the bodily systems to reintegrate. This therapy is very different from current speech therapy practices and remedial methods.

Speech therapy provides only compensatory corrections that actually take away part of an individual's IQ, thus further hindering learning abilities. In fact, children identified as speech handicapped do a little better *without* correction by present-day therapy programs.

There is nothing bad about the programs. They simply do not treat the cause of speech and learning problems; instead they treat the symptoms. This leaves children with deficits. Remember Jane in chapter three? She had to think how to speak. She used thinking ability to control her speech as she was taught by conventional speech therapy methods.

I am not blaming the speech profession, because the cause of speech disorders was not known when the field came into being. Because of the low success rate with speech therapy's compensatory methods, however, school policies now *prohibit* early intervention of speech disorders. This is due in part to the belief that speech is developmental or learned. Speech is *not* a developmental or learned skill.

I have discovered that speech is a motor skill that can be correctly trained before or after birth by the proper thumb suck. This explains why many children of all nationalities speak clearly when they start talking. They do not have to correct any of their sounds on their own or with help.

The struggling children who can be helped with my therapy instead are victims of school policies and politics. They are in a position doomed to failure except for those with extremely high intelligence who can compensate for their own deficiencies. In fact, most speech pathologists are becoming language teachers. They have given up the real mission of finding the cause of speech disorders to bring about a symptom-free child.

Is there any hope? Yes. If these children could have the cause of their problems identified and be allowed to go through my therapy, they can be cured. These children need to go back to where their problems began so that their neural pathways and muscle reflexes can be reintegrated and retrained—so that they no longer have to think to control deviant speech and learning patterns. Let's look at another case as an example.

Joey's Turning Point

Joey, a little Hispanic boy, had difficulty making *t, d, n, l* and *r* sounds. His *ch* and *sh* sounds were distorted, and the *s* and *z* sounds were not well made.

Joey started public school kindergarten and made his way through it nicely. He was very cute—the teacher's pet. Then he went into first grade where I met him during a screening. He entered a speech therapy program much like those still going on today. Joey spent the entire first-grade year having little success. I talked with the kindergarten teacher who described Joey as a vivacious, outgoing child. Joey as a first-grader could not be recognized by that description.

I was developing my new therapy program at this time and knew that Joey was ideal for it. His parents agreed to allow me to work with him using the new therapy. I had Joey work with his thumb in his mouth. In about three months, he started to blossom. In fact, his teacher reported that the quiet, little sedate Joey had started to act out.

One day Joey decided that the classroom was hot and the windows needed to be opened. The classroom monitor had not done the job, so Joey got up from his desk, picked up the window pole, and started walking toward the windows. The teacher shouted, "Joey, what are you doing?" He turned around quickly and smashed a ceiling-light globe, sending glass fragments about the room. Fortunately no one was cut.

Even though Joey could now talk clearly, he had learned to work nonverbally. He had quit talking when others told him he talked funny. Immediately following the globe event, I reminded Joey that he could speak correctly now and needed to talk to people before he did such things. He learned through the consequences of his breaking the globe.

Joey's teacher wanted to know what happened to her quiet student. She had not met the real Joey, the Joey who was not afraid to talk, the Joey who was now cured of his learning problems. The true Joey became a delight to teach.

What caused his problems? He sucked on the middle two fingers of his right hand. As he did this, the index and little fingers were pressed into his face, causing little scar-like indents under his eyes. His parents always thought these were from cat scratches or something similar.

The two-finger suck habit likely started in the womb. This habit created his speech problem and consequent learning problems. Joey experienced tremendous emotional release when he started to act things out. I counseled him about talking things out instead of acting out. He understood this and adjusted.

Joey's speech sounds corrected beautifully as he used his thumb to retrain his motor reflexes. After his finger habit was stopped, a dental malocclusion corrected itself, serving as evidence that Joey indeed had given up the habit. Such structural changes occur within three to four weeks. The scars under Joey's eyes cleared up, and he acted more like the child described by his kindergarten teacher. Joey showed gains in his classroom work as therapy progressed. The difference in him was like night and day. Joey was on the right track.

The emotional release Joey experienced in the course of this therapy marked a turning point for both him and me. It allowed Joey to express his feelings and desires. For me, his therapy results clarified what was happening to my other young patients. I had been getting feedback from parents and teachers about behavior changes. This had not occurred with traditional speech therapy methods.

Fifth Gwade Child in Twouble

I had the opportunity to work with a fifth-grade student shortly after I started using my new therapy program. Jonathan moved to my school district in Los Angeles from

Chicago, where he had been in public school speech therapy as well as a private speech clinic. Because my school screened all new students, I found Jonathan and offered my therapy to him.

Jonathan's records showed that he had trouble with the *t, d, n, l,* and *r* sounds. He was one of the most pronounced *w* for *r* children I had ever met. I wanted to place him in my program, but his teacher strongly opposed speech therapy. She was very outspoken about not seeing results from speech therapy programs during her many years of teaching. She also served as a board member of the school district, so she wielded influence on important decisions.

When I approached Jonathan's teacher, she demanded to know: "Why do you want to take this boy out of my class for therapy?" I explained that I had a new therapy to try that might help Jonathan with his speech problems as well as other problems. I asked for a three-week trial. Because she was not pleased with Jonathan's schoolwork, the teacher agreed.

Jonathan had problems in math. They were doing word problems, and the teacher could not even get him to read in class.

I found out that Jonathan put an index finger way back into his mouth to go to sleep at night. A callous on the finger provided evidence. The finger held the tongue way down, so it could not produce the sounds made at the top of the mouth (*t, d, n, l,* and *r*). This habit also caused distortion of the *s, z, sh, ch, zh,* and *j* sounds.

As I started to work with Jonathan, I noticed that he wasn't really trying. I asked him why. "If I change the way I talk, people will notice and make fun of me," he said. We discussed the fact that other kids were not making fun of him any more for the way he talked.

I gave him a challenge. I suggested that Jonathan (who was very popular and excellent in sports) wear unmatched socks to school the next day and tell me how many people noticed. We obtained permission for him to go home from school early if classmates teased him.

The next time I saw Jonathan, he was waiting for me at my office door in a very excited state. He said, "I wou uh bewieve ih! I wou uh bewieve ih! I wou uh bewieve ih!" He sounded like Porky Pig. I asked, "You wouldn't have believed what, Jonathan?"

No one had noticed the unmatched socks. Jonathan even wore the other unmatched set a second day, and still no one noticed. I breathed a sigh of relief! Now Jonathan was ready to work for me.

First I asked Jonathan to get the finger out of his mouth. He asked how I knew about it. I told him that I felt a callous when I examined his hands. Within two weeks, Jonathan was able to make the *t, d, n,* and *l* sounds. The *r* took only one more week. In this short time, Jonathan's classroom skills picked up remarkably. The difference shocked his teacher. She made a special trip to my office to inform me of all the changes that were taking place.

I saw Jonathan for two months, and this included diagnostic sessions. We worked with the thumb to reestablish motor reflex patterns for reading and spelling as well as speech. His teacher showed me pre-therapy papers in which Jonathan misspelled words like he mispronounced them. She showed me post-therapy papers in which he spelled words correctly.

Jonathan came out of the womb with a fairly high intellect and a habit that gave him great difficulty coping with school subjects. I met his mother shortly after dismissing him from therapy. She was amazed at the changes and said, "Thank you for giving me my son back!" I asked her what she meant. She answered with tears.

"My husband was so ashamed of our son that he would not take Jonathan anywhere. This has to be an act of God, because we are only here for a short time; in fact, just two more weeks. My husband was on loan to a company here. We are going back to Chicago. I have been praying for something to happen. My husband just took Jonathan to a ballgame for the first time!"

One, Two, Twee, Four, Five, Sex

I found that my therapy worked for adults with speech and learning problems as well as it did for children. One of my first patients, an airline stewardess, was about to lose her job. Her responsibilities included talking to passengers over the microphone. This made her nervous, thereby causing a return of her infantile speech pattern that she could control in most situations. I had her count to ten with her thumb in her mouth. It brought out the old pattern.

"One, two, twee, four, five, sex, seven, ape, nine, ten." The stewardess was stunned as she heard herself. "No wonder people laugh at me," she said.

Therapy was the same for her as it was for others. We worked with the thumb to retrain her motor reflex pattern. She corrected her infantile pattern. She kept her job and lost her fear of being laughed at.

Not Child's Play

Many people with whom I've worked, young and old, first think that practicing with their thumb is child's play. I get the most cooperation from those who are in pain or who fear losing their jobs. They will try anything. In fact, they would suck on their big toes if that's what it would take to get rid of their problems.

The only sad part of therapy is that many patients will not tell others how they solved their problems. They refer others to me, but they don't explain what I ask patients to do.

We are on a road to learning all our lives. I hope one thing I can contribute to our learning is that the thumb is more crucial to our well-being than we've ever realized. It is the key to start us on the right road; it is the key to getting us back on track when we journey off. By knowing how the right road to learning is marked, we can avoid the bumpy roads. For those of us who missed the right road in the beginning or veered off at some point, we can turn back, get on the right road, and catch up to our peers. There are no speed limit or age limit signs to prevent our restoration.

With the thumb we can hitch a ride back in time and pick up the right path.

Part Two

The Pebble in the Shoe (Finding Causes)

Chapter Five

The Pebble, Speech, and Dentistry

The way an individual speaks, his articulation, says much about him. For the trained examiner, how a person moves the jaw and tongue while talking provides clues to problems with jaw and ear function, language and learning processes, speech and voice production, and the dentition. The articulation also tells much about the causes of tension-pain patterns and facial growth imbalances.

Articulation is crucial to diagnosing. A defective speech pattern is a symptom of the cause of not only the poor pattern itself but of other problems connected to the oral structure. Articulation is crucial to the diagnosis of problems addressed in this book.

But old articulation therapy—what I used to do, what speech pathologists still do, what many parents and teachers do, and what many individuals do to themselves—does not treat the cause of the defective speech pattern. The old therapy was derived from seeing a problem (an unacceptable speech pattern) but not understanding that the problem was a symptom of something else taking place. This led the speech therapy profession into a symptom-chasing practice. Therapists concentrated on the symptom and finding ways to compensate for it.

As I reviewed material in the speech field, I found that researchers only addressed possible causes when they thought birth trauma or forceps delivery might cause articulation problems. Their research did not reveal a cause.

I saw that researchers had overlooked one important fact: Some children start out talking beautifully, the articulation is there, everything is fine—then for some reason they wake up one morning and cannot put two words together to make any sense. In such cases, the disorganization is not in the language but in the articulation of the language. These children start struggling to put things back together the right way. These cases have been many, but no research has been done on their behalf. Instead their articulation defects are merely described by therapists who put them into subgroups according to what sounds they cannot make correctly. Then they are introduced to various ways—compensatory ways—to produce the sounds acceptably.

But many fail to produce acceptable sounds and are written off after years of therapy as having met their potential or not willing to work hard enough.

Speech pathologists describe an articulation pattern by listing what sounds are omitted, distorted, or substituted with other sounds. They are not trained to address causes.

I often illustrate the difference between describing symptoms and identifying causes by presenting a made-up scenario. It's about a missed pebble.

The Pebble in the Shoe

Imagine putting a very sharp pebble in someone's shoe. This person walks normally and has never limped. Suppose this person does not realize that there is a pebble in his shoe. He simply feels a sharp pain in his foot and adjusts the way he walks to alleviate the pain. The adjustment takes care of the immediate pain, but later on other problems present themselves. Sometime down the road this person seeks professional help for his problems.

An examiner sees that the individual walks on the outside edge of one foot. This causes him to limp with the ankle, knee, and hip turned outward. We could have placed the pebble in a different part of the shoe to make the foot turn inward. Or perhaps to the front of the shoe to make him put most of the pressure on the heel. Or perhaps to the back to make him stay off the heel. Regardless of where the pebble is placed, it can cause a set of problems to develop.

An examiner will very likely describe what he sees—how the person limps. And he usually attempts some kind of logic. "This person walks on his instep rather than moving the foot in a heel-to-toe movement, causing him to limp and to . . ." The only true logic to the whole scenario, however, is that there was a pebble in one shoe—a pebble placed in a certain position.

The Pebble and Speech

Speech therapy for articulation disorders is handled much this way. The symptom (speech) is described and treated. The cause is not addressed. Instead the evaluation might read: "Johnny does not make the *s*-sound correctly. He sticks his tongue out, causing the *s*-sound to be substituted with a *th*-sound." His case is tagged "frontal lisp." Johnny is placed in a subgroup of articulation cases. Children with similar speech symptoms are often placed together in a public school therapy program to work on compensating for their problems.

I believe that putting children into subgroups for speech and learning problems really shortchanges them. These children are being lumped into groups for funding purposes. Since therapists treat only their symptoms, many children remain in the groups for a painfully long period often with little or no success.

Seeing stutterers placed together in a group really disheartens me. I have found out that stuttering is an articulation problem, and the causes for articulation problems are quite diverse. Yet they have one thing in common: Something is placed into, up against, or around the oral structure that hinders correct function. (Think of the pebble in the shoe.)

To people who don't believe that oral habits cause speech disorders, I have given a "pebble in the mouth" challenge. I say, "Give me an adult for about a week—someone who will do what I ask. I'll have that person lose control of his speech, making errors he's never made before." Actually a teacher challenged me one time on this idea. She said, "No, that's not possible." I offered her the chance to experiment. She accepted the challenge and asked what she'd have to do.

I showed her how to place two fingers in her mouth in a particular way. I told her to repeat "See Sammy Snake swim" several times with the fingers inserted. After several minutes of practice, she then asked, "Now what?"

I said, "That's it. That is all you have to do. Sometime today or tomorrow you'll go to say something, and it will come out the wrong way."

I met this teacher again at school and began a conversation with her. She first told me that nothing happened. I responded, "Sometimes you win, sometimes you lose." She then said, "No, I'm only fooling. When I went home that night, the space heater in my apartment wasn't working. When the repairman entered my apartment to fix it, he asked, 'Where is it?' I said, "My thpace heater ith over there.'"

The teacher then said, "I have never done that before in my life!" I responded again, "Win some, lose some."

Causes, Not Symptoms

What happens in speech therapy and other health-care fields, as well as in education, is that observations are made, then assumptions are made. Once assumptions are made, validity with the observations is lost. To be valid with observations requires finding a direct linkage. But we seem to have started looking at symptoms as problems to be treated and have neglected the scientific approach that would simply say: "We have a symptom. *Why* do we have this symptom?"

Most of our therapies or treatments by many fields today tend to be pain oriented. A person says, "I don't care what the cause is, get rid of the pain. I can't stand the pain," so the pain, not the cause, gets treated. But by getting rid of pain, many times we miss the avenue where we could go back and pick up the cause of the pain. So the pain becomes a reoccurring condition.

Early in my career when I asked professors how speech problems begin, the overall conclusion was that "speech problems are just there." In talking to different leaders in the field, I learned that they believed that we don't have to worry when a speech problem started—we are to be concerned about the current condition. They did not see that what happened a long time ago had any value in diagnosing. I disagreed with them.

I believed from the beginning, and now am sure, that what causes a speech problem is what happens right on through an individual's whole pattern of growth and function. We must go back to that happening's beginning, the cause, and stop it to get rid of the symptoms.

But in the field of education, symptoms are treated as different problems, so there are different therapies or lessons for different groups. This is a disservice and has led to a great injustice in the schools. Consider those who stutter as an example.

Stuttering is no more than an articulation problem. This may sound overly simplified, but the individual is in the wrong production position to make the sound that he wants to say. Some people say they have observed that stutterers can at times say the words they want to say, or they can repeat words after another says them, or they can talk to pets without stuttering. The reason for this is simple. In these instances, the speech has not been thrown into the reflexive pattern. Get the stutterers into their reflexive, uncontrolled pattern and they will stutter regardless of what they are doing.

The treatment of stuttering and other articulation patterns are being delayed in the schools with the hope that by waiting long enough the children will grow out of the problem. But as I have said in earlier chapters, the causes need to be found and eliminated as early as possible to prevent compounding problems—problems in language and learning development and dental growth.

The Pebble and Dentistry

Now let's take a look at the pebble in dentistry. We have established that oral habits cause dental as well as speech problems. Dental abnormalities can exist without speech problems, however. It all depends on how the oral habit is used.

I've made the point that speech therapy has developed as a symptom-treating profession. In many ways, this is true of dentistry.

A drastic type of symptom treatment is oral surgery. It is sometimes used to reconstruct a jaw that has grown forward because of an oral habit. One interesting case I had involved a child who was growing her jaw forward and her mother who wanted to avoid this surgery at all costs.

She was only four years old when I met her. I'll call her Susie. Anyone who looked at Susie wondered what was wrong. She looked angry all the time.

Susie's mother worried about her daughter's jaw growing forward; it caused the upper teeth to be positioned behind the lower teeth, giving Susie a mouth that resembled a bulldog's.

Susie's mother, a surgical nurse, had assisted on a number of jaw surgeries. When she realized that her daughter's jaw was growing into a position that might require surgery, she visited her friend Dr. Berke. I happened to be in his office at the time.

Dr. Berke told the mother that orthodontics did not benefit such a young child; typically nothing was done at such a young age. They'd just have to watch the growth to see what happened.

Thinking that her daughter might face the surgical procedure devastated the mother. Dr. Berke called me in and asked if I could do anything to help. I agreed to try. I requested a panoramic X-ray, some side-view pictures, and a set of molds on the child's teeth. While Dr. Berke visited with the child, I talked with the mother.

I said, "When I first saw your daughter, I wanted to ask her what was wrong. She looks angry. Yet I bet she would say that everything is fine. I have a feeling that she is growing this way because she sticks her jaw out." The mother didn't suspect this possibility, so I asked her, "What do you do when you get mad? Give me your most ferocious look." She did. Where did her jaw go? Way out!

"Oh, my God," she said, "I am teaching my daughter to grow that way! And she is already there!"

I said, "I know, but if we can keep her from throwing her jaw forward—not let her do it, and if you will watch her and yourself, we can probably correct it."

We started into therapy. It took only three months to get that little girl's top teeth over her bottom teeth.

About four years later Susie and her mother visited the office to show us a beautiful set of natural teeth that gave no indication of the bite that she had at age four. The mother expressed her thanks and said, "Even to this day when I mop the floor, am concentrating on something, or doing any number of other things, I still have a tendency to throw my jaw forward. But I catch myself and just don't do it."

Simple observation and simple instructions saved this individual from complex and chancy treatment to correct a bulldog appearance. I simply showed this young child how she needed to bite off her food and how to chew it. I showed her where she needed to keep her teeth. Mother Nature did the rest—she came in and put that child's mouth into the shape that it was intended to have.

Extractions v. Nonextractions

Dr. Berke was the first dentist to notice that my therapy to eliminate oral habits and to teach correct function of the oral structure resulted in a natural correction of dental conditions—particularly arch formation and the alignment of teeth. Together we did an in-depth study of dental-related conditions.

Dr. Berke had expressed much displeasure about his profession; he suspected that some treatments caused more damage than good. He wanted to watch what I was doing and support me with dental evidence. We made molds, took pictures, and documented changes. He taught me much about dentistry, and I shared my discoveries with him.

Dr. Berke was concerned about young orthodontic patients who came into his office for extractions (pulling of teeth), particularly of the permanent bicuspids, or serial extractions of the lower front four baby teeth. He thought that extracting the front baby teeth almost guaranteed that the permanent teeth would come in crowded.

Many in the dental profession think they make space for the permanent teeth by extracting the baby teeth. Dr. Berke and I saw that the opposite is true. By removing baby teeth, a child is being maimed. Space is lost. Growth is lost. Bone loss occurs when teeth are extracted. That bone loss keeps children from growing the way they naturally would.

Dr. Berke and I found that we could identify what the children were doing to cause the crowding of their teeth, help them stop their habits, then teach them to use their mouths correctly. This brought about nice, natural corrections without removing teeth.

We spent nearly six years compiling information with models and pictures to support our results. Some of the pictures are found in chapter nine.

Once we had our work well documented, we started sharing it with orthodontists, hoping they would be pleased with the discovery of causes. Instead they became upset and did not want to hear about it. We soon realized that the information would make them liable. If their patients found out that they knew what caused the problems and did nothing about it or refused to go along with better treatment plans, they could get in trouble.

Not all orthodontists prescribe tooth extractions. In fact, orthodontists can be put into one of two groups: the extractionists and the nonextractionists—the tooth pullers or non-tooth pullers.

An extractionist tells a person that he has too many teeth for his mouth or that the teeth are too big—"perhaps you have your mother's jaw and your father's teeth or vice versa." But nature, our design, does not mess up like that. To pull teeth makes the orthodontist's job of straightening the front teeth much easier—never mind that the bite may be off and the ability to chew food properly has been hindered!

The nonextractionists do experience some failure in keeping the teeth straight after braces are removed. But the reason for this is not poor dentistry. Most of these orthodontia cases return to their pretreatment condition because the same oral habits that created their problems are still active. When teeth are not removed, the original position can be reestablished. Therefore, nonextraction cases have to wear retainers for years after the braces are removed.

An example of this is a soldier who lost his retainer in Vietnam. He wrote to his dentist asking for another one to be made and mailed to him. By the time he got the replacement retainer three months later, it did not fit. His teeth had shifted that quickly.

A ten-year study was conducted by a German orthodontist, Dr. Claus W. Schwarze, who confirmed that once a retainer is removed, whether it be six months or ten years after the completion of orthodontic work, the teeth go back into their original position. Orthodontists came up with the idea of extracting teeth to help prevent this reoccurrence. They decided that the mouth could be changed, shrunk back on the ridge, by extracting teeth. This was done with head gear, using the molars as an anchorage point for pulling the teeth back on the ridge.

Eight permanent teeth are often extracted with this orthodontic procedure. To get an idea of how much the mouth is shrunken, imagine eight teeth and their supporting bone tissue placed in the palm of your hand. Generally, eight of the larger teeth, four molars and four bicuspids, are pulled to guarantee that your mouth will not return to its preorthodontic state. Eight teeth make quite a pile in the hand. Not only is the mouth shrunken by that amount, but the bone that surrounded those teeth is lost. Anyone can see this by looking at dental X-rays taken after teeth have been pulled. What does this do to the facial structure? A British study gives the best answer.

The Good, the Bad, and the Ugly

Twin girls in England had orthodontic work done. One girl was treated by the nonextraction method; the other by the extraction method. As they grew, the one whose teeth had been pulled wanted to know why her sister was growing so pretty and she was growing so ugly. These girls had everything else in common, but their facial growth patterns did not remain the same. The sister who had the extractions refused any further follow-up after several years. She refused to see or talk to the investigators. She had become emotionally distraught.[1]

Those who conducted the study claimed that this was not what they expected; they even denied that they were doing a comparative study. But I find it hard to believe that they had anything else in mind. Otherwise the treatment for both girls would have been the same.

When four or eight teeth are extracted and the remaining teeth are pushed back on the ridges, the mouth cannot continue growing the way it was intended to grow. Pulling the teeth back at an early age retards bone growth. The third molars usually become impacted, and other problems with the oral structure often develop.

All this can be avoided. When children receive therapy to get their habits stopped and to recondition their system for correct function, the results are far better corrections than anything orthodontia can provide.

I have seen some sad cases in which children actually worked to destroy their teeth so that they could wear bands and braces like the "in" crowd. I question the value of orth-

1. "An Orthodontic Challenge" by H. L. Eirew, B.D.S., L.D.S. *International Journal of Orthodontics*, Vol. 14, no. 4, Dec. 1976

odontia from the problems and the permanent damage I've seen in these cases.

We are one of the few societies that is so set on false teeth. Most people do not understand that our Hollywood actors and actresses usually have all their teeth ground down to pegs, then capped. And for teeth that have been replaced, that beautiful smile can cost up to $100,000.

There is nothing prettier than teeth that come in naturally to the proper bite position. But many times parents look to orthodontic treatment to make those front eight teeth straight. They are oblivious to what goes on in the back of the mouth—how well food can be chewed and swallowed.

Unfortunately putting braces on the teeth to force them into more desirable position does not treat the root of the problem.

In fact, the roots of teeth become diminished due to the battle of forces between active habits and opposing arch wire forces. This is why many people complain about their teeth feeling loose. Some experience teeth that go into periodontic (gum tissue) distress; some experience loss of teeth.

The continuing rocking activity causes tooth root resorption (disintegration), causing the tooth to lose its anchorage. This can be compared to the rocking motion carried out by a child assisting the removal of a baby tooth.

There is a limit to what the tooth can take. I call it the "50 percent margin." Past that margin, the chances of saving the tooth are almost nil. Surgery or bone grafts sometimes work to save a tooth.

Removing the Causes

Orthodontic braces are not a cure. Speech therapy to change speech with compensatory activity is not a cure. Instead other problems are created when symptoms, not causes, are treated.

Think about the pebble-in-the-shoe illustration. What if the pebble was positioned at the front of the shoe, or the outside edge, or at the heel? How would the symptoms differ? How would the body modify itself to live with the pebble? How do the jaw and the teeth adapt to accommodate an oral habit? It is the same principle—a direct link, a cause-effect relationship exists.

The key to curing is to identify and eliminate the cause. The pebble must be found and removed to achieve a cure.

Cute Habit?

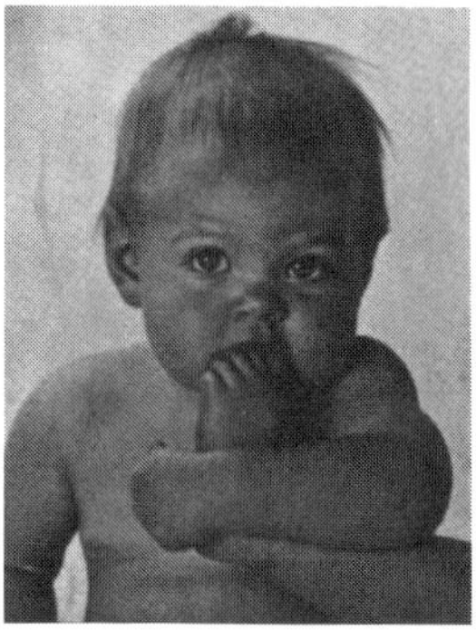

Very cute! Oh, but how potentially damaging! This child's habit of putting the left foot in his mouth and sucking the toes may be cute, but can you imagine all the damage he could do to himself if he continues this habit for several years?

This child is starting to grow his face crooked, and it appears that the ankle and foot are becoming deformed. Continuing the habit will produce an open bite, probably growing into a crossbite. It will cause deformity in the ankle, foot, knee, hip, toes, shoulder, upper back, and wrist, and could cause the leg to bow. He will have difficulty walking. This habit will produce a severe speech problem leading to language and learning problems, especially in reading and spelling. The child will develop biting, chewing, and swallowing problems, which can lead to indigestion, pre-ulcerous conditions, and colitis.

Remember all of these problems have the same cause—the oral habit. Yet how many people would even consider this to be an oral habit? How many people would ever relate the problems to a single cause much less to an oral habit? Think of how many different practitioners would try to solve these individual problems without looking at the whole picture. They would try to treat the symptoms without getting rid of the cause. Unless the cause is eliminated, successful treatment of any of these problems is impossible.

The best treatment considers the whole picture.

Would You Believe?

In 1980, the conclusions below were drawn from Streicher's research. He suspects some of these statistics are higher now.

- At least 85 percent of all orthodontic patients are nail-biters or former nail-biters.

- Over 75 percent of the patients seen for TMJ (jaw joint) problems are nail-biters or have habits similar to nail biting.

- At least 85 percent of all adults seen in pain clinics are nail-biters or former orthodontic patients who have had four bicuspids and usually four wisdom teeth extracted.

- The leading cause of delayed speech is a large amount of cloth being shoved into the child's mouth, which completely overwhelms the natural feelings in the mouth and makes it physically impossible for the child to produce intelligible speech.

- More than half of the patients seen by periodontists are nail-biters.

Chapter Six

The Pebble, Psychology, and Medicine

A dentist presented a startling observation not long after I started treating patients in his office. (I worked in several dental clinics.) He said that he was seeing more emotional changes in his patients as a result of my therapy than he had ever seen in patients who were receiving direct therapy by psychologists and psychiatrists.

I thought, *Wow! Something is going on here that is uncorking a lot of emotions.* I was aware of emotional release in my school students, but now adults were also experiencing emotional release. I became greatly concerned about this when my first six stuttering cases ended in divorce. I decided to investigate.

I called two of the former stutterers and two psychiatrists—one who favored Freudian theory and another who was Jungian oriented. I first visited with the stutterers. I'll call them Joe and Mike.

"Joe, what happened to you and your wife?" I asked.

He said, "We were out at this restaurant. My wife had always done the ordering for me because I used to make an idiot of myself trying to place an order or even to ask for a cup of coffee. Well, this time she placed her order, then I spoke up and told the waitress what I wanted. My wife sat there with this look on her face that you just wouldn't believe. She said, 'What is the matter? Don't you like the way I order your food anymore?' I was taken back by this. The waitress came with the coffee. It was lukewarm, so I asked her to bring another cup and to make sure that it was hot. By this time my wife was really hot."

Joe explained that his wife basically had been running his life, even making all his phone calls for him. She did not see his ability to do things for himself as a positive change. The change did not fit their usual lifestyle.

I called Mike. He said, "About the only thing I can tell you is that one evening my wife and I were sitting in the living room. The phone rang, and I got up to answer it. It was a wrong number, so I hung up fairly quickly. My wife looked at me with fire in her eyes and said, 'Who was that that I wasn't supposed to hear?'"

Mike's wife was not excited about his ability to answer the phone. I thought this wasn't much of a relationship. Here Mike used to stutter so badly that he wouldn't go near a phone. He would not sit in a room alone if there was a phone in it. He would not stay alone near a phone at work.

I took my observations to the psychiatrists for their assessment of what was going on with my therapy program. We discussed some case histories, and I asked them if any damage was being done by my therapy. They said no. All the changes I was seeing were positive changes—even the divorces. They assured me that I wasn't doing anything but making people better, that I was giving people more insight and making them operate better in their own environment.

From that time on, I set up my therapy to include spouses or special companions so that they could heal at the same time. Therapy for children included parents. I warned parents of possible emotional release. I encouraged them to put away any valuables that the children knew not to touch, because those things could be involved in an accident during the therapy. Once we got past the point when emotions would release, then the valuables would be safe again.

Now I want to toss the pebble into the field of medicine.

The *Art* or *Science* of Medicine?

The ingestion of chemicals and dyes from fabrics by many children, the current treatments of jaw complaints, and the overall scheme of treating symptoms instead of pursuing causes are medical issues that greatly concern me.

I have watched for years for any research being conducted on such chemicals and dyes ingested by children. I am not aware of any, but I believe research is warranted. I have treated many children for cloth habits. Much of the material that they suck goes through washing machines with various cleaning chemicals and detergents. These machines do not get rid of all the residue, so children must be taking it into their systems. They suck on stuffed toys, blankets, clothes, washcloths, and other items day and night, night and day. Recall the little Japanese girl with the green teeth. How much dye did she ingest? Does she suffer from stomach disorders or colon cancer today?

I'm concerned because this issue needs scientific research, and our doctors today no longer say they are practicing the *science* of medicine. They have changed it to the

movement changed their ear canal pressure. To get a feel for this, you can stick your little fingers in your ears and start moving your jaw. If you form a seal in the ear canal, you will not only feel the seal give way as you rotate your jaw or move it side to side, but you will feel the pressure changes against your fingers.

Hearing aid molds pop out of the ears with the change in pressure created by incorrect jaw movement. This was a very embarrassing situation for one woman whose hearing-aid mold suddenly popped out, causing very loud feedback that attracted stares from other diners while she was at a restaurant with her date.

I taught these individuals how to use their jaws to get the minimum amount of ear mold rejection so that the ear mold would stay in place. I taught them how to bite straight up and down, how to chew the same amount of food on both sides of their mouths to allow the food to move back to the molar area, and to keep chewing until a swallow occurred without them thinking about it. I saw this as a readily correctable problem and thought nothing of it as a specialized practice. Not until I got a threatening letter, that is.

A California state licensing department informed me that I was being accused of practicing TMJ without a license. The complaint was filed by people being protective of their surgery business. The irony is that people who have had surgery on their jaw joints really need my therapy. But some TMJ moneymakers realized that my therapy prevented the need for surgery. Instead of visiting with me to find out about my work, they chose to attack. Perhaps they started out really wanting to help people, but now they had to protect their pocketbooks. I was investigated and found to be doing nothing that they could legally complain about.

In working more and more with TMJ cases (some prefer a later tag like TMJ syndrome or TMD—temporomandibular disorder), I discovered that many people had been to their dentists and physicians only to find out that nothing was wrong physically; that the problem must be in their heads—an emotional or psychological problem. My wife's grandmother was one of these people. I regret that I was not far enough along in my work to know how to help her. I can look back now, as I do with Aunt Hattie, and see how much unnecessary suffering Grandma experienced.

Grandma's dentist pulled every one of her teeth and gave her false teeth that never really fit her properly. She always had trouble with them and was pegged as a complainer. She suffered headaches and was kept on drugs twenty-four hours per day to prevent her from hurting. I regret that she died before I did my research. Now I know that she did not need to lose her teeth, suffer the pain, or be medicated half out of her mind. She could have enjoyed life.

Now any jaw discomfort or pain is called TMJ or TMD, but basically it is the misuse of the oral structure—a simple problem that is often accompanied by migraine-like headaches. These patterns are not difficult to correct. It simply requires teaching individuals to relax—not just telling them that they need to learn to relax. Many people are being misled into believing that running, kickboxing, or doing jumping jacks to get aggression out are ways to relax. But getting physically active is not the equivalent of relaxing.

As I already have pointed out, the muscle tension needs to be released and disconnected from external stress.

Watch the Talk

I tell the people I've trained: "Watch a person speak and you will have your first clue, an obvious symptom, to what the root cause of the complaint is. Teach that person to use the oral structure correctly, to speak without a jaw shift, to bite and chew properly, and to stop the oral habits causing the dysfunction, and you'll have the individual back to where nature intended him to be."

Does this therapy take care of everyone's problems? No. But I tell you that people who go through this therapy are so much better equipped to cope with life. The therapy is not just for cases seen by the speech field but by those in education, dentistry, psychology, and medicine.

Predetermined Ideas

Professional fields such as medicine, dentistry, psychology, and speech all started out much the same way—with predetermined ideas. There were things people wanted to prove.

I did not set out to prove anything. I spent years in college only to find out that no one could tell me what causes speech problems. I had to listen to assumptions, such as "a kid sticks his tongue out on the *s*-sound because he is mad at his parents."

I see most of these fields as locked into a regimented, predetermined way of operating. They even have predetermined ideas of where new ideas will originate. I have run into many unbelievable reactions to my research and clinical results. I'll tell one of my favorite examples.

My mentor Dr. Berke invited a friend of his to the dental office to see our original research. The friend was a neurosurgeon from one of the leading universities in California. Dr. Berke asked me to get out the models and pictures that supported our work. This medical doctor looked over all the evidence, then turned around and said, "If this stuff were true, I'd know about it." Dr. Berke's mouth fell open. He had just shown his friend original, well-documented research and proof of successful treatment. The message from Dr. On-Top-of-it-All seemed to be, "Who do you people think you are? How could you know something I don't?"

Well, I guess we are simple people—willing to open our eyes or to ask God to open them so we can see what is going on around us.

Chapter Seven

The Pebble and Education

Jimmy, four and a half years old, could not talk. His mother was distraught when she brought him to my office. Two speech pathologists had worked with Jimmy and couldn't help him. School officials believed he was retarded and would not consider him for enrollment.

When I met Jimmy, I saw evidence of a cloth habit. His teeth showed cloth wear. He had excessive saliva. His lips were tightly closed, and he had a suck-swallow pattern.

His mother confirmed the cloth habit. "He even drinks his water out of a washcloth," she said.

I told Jimmy, "Your shoe is untied." He glanced down at his shoes, which were tied securely, then he gave me a look that said, "What's wrong with you?" This suggested that his language was quite well established.

I spent a couple of therapy sessions trying to get acquainted with Jimmy. I needed to break down his defenses. During his third visit I sat down on the floor by the door and told him that he couldn't leave until he told me to open the door. It took only five minutes for him to say, "Door." I asked him to say, "Open the door." Jimmy repeated the phrase, but even he knew the articulation was poor.

I asked Jimmy to show me how he chewed on cloth. He took his T-shirt, formed a tail from the bottom of it—about six inches long—and shoved it into his mouth. He pulled up more of the shirt, then more and more until it crept up around his armpits. I wondered how he didn't choke. I asked Jimmy to repeat "Open the door." His speech had a nasal quality, but I could understand it. This confirmed that the cloth habit was responsible for the unintelligible speech; he had conditioned his mouth to function with cloth. Without cloth in his mouth, Jimmy's speech could not be understood.

I told Jimmy that his chewing and sucking on cloth had to stop if he ever wanted to sound like other boys and girls. As therapy progressed, I talked to Jimmy about various types of cloth. He reported that T-shirts and sweaters taste really sweet. He made an awful face at the mention of curtains. He knew the taste of a variety of fabrics.

Jimmy stopped his habits and completed therapy, conditioning in the speech pattern that he would have had if nothing had impeded it. He entered school and did fine. His mother stopped by my office sometime later to tell me how happy she was and to jokingly ask what she thought she would never ask: "How do I shut him up?"

For Jimmy, the pebble in the shoe was cloth. Had the "pebble" not been found, Jimmy may have started school classified as delayed in speech and mentally retarded. His problems would have become more and more complex. Present day remedial methods would have complicated his problems. Jimmy, and all children with speech disorders, need to find the cause of their individual problems. It is necessary to determine how the learning path was established by going back to its physiological basis, then establishing the intended path.

In Jimmy's case, we removed the cause of the problem and introduced a new speech pattern. I did not put his speech into a pattern that was foreign to him. The correct pattern was the pattern that would have been his speech without the cloth impediment. If he had imitated somebody else's speech, if he had learned to watch and copy what another person did in speaking (a technique used in speech therapy programs), his speech would be unnatural for him. He would be controlling his speech by using another person's speech pattern. This would suppress his learning and reading skills and reduce his intellectual competence.

The speech pattern that appears corrected but is not internalized can be compared to a foreign language. Because the defective pattern remains internalized, the controlled speaker has to decode what he hears spoken, encode it within his own system, then decode and encode again in order to speak or read. This process calls upon cortical (intellectual) ability. The thinking part of the brain works to control how words are spoken. Only the very intelligent person can concentrate on his message or comprehend what he's reading as he speaks or reads. But to do this reduces his potential.

Therefore reading comprehension is adversely affected by teaching a person to control speech instead of finding the cause of a speech problem and reconditioning the speech pattern. One young woman immediately comes to mind.

Heidi and Sherlock Holmes

Heidi, a college student, came to my office when she was failing a required political science course for the second time. It meant Heidi would flunk out of school. Her mother, a remedial reading teacher, couldn't help her.

"What do you like to read?" I asked Heidi.

"Sherlock Holmes," she said.

I started discussing events in the Holmes mysteries, but Heidi could not recall what she had read. She had difficulties remembering what she read whether it was something that she liked or didn't like.

As a young child, Heidi had received speech therapy. She still had difficulty controlling her production of the *t*, *d*, *n*, *l*, and *r* sounds. Thus, her speech pattern was controlled; it did not flow easily. A compulsive reader, Heidi chose to read books rather than to go out on dates. I asked her how many times she had read her political science book. "At least seven times," she replied. "And I've spent hours studying for my tests."

After working for a while to recondition her speech, I had Heidi read some Sherlock Holmes. All of a sudden, she stopped and blurted out, "That black-haired witch! She would not let me point with my finger!"

That black-haired witch had been Heidi's second-grade teacher, who had decided Heidi should stop pointing with her finger as she read. It was not an unusual request by the teacher. Reading one word at a time causes the reading to be choppy and hinders comprehension. But the teacher did not know that Heidi needed the extra time with words to process them fully. She had to go through the decode/encode process twice in order to read the words out loud correctly.

Many children continue to read words one at a time, but they learn to do it fast enough to sound smooth. They do not have the foggiest idea of what they are reading, however.

My Diagnostic Scale

In my research I placed speech deviations on a scale of one to six. Six stood for a nice, flowing, incorporated speech pattern. In testing 300 school children, I found that only four or five students fit this highest rating. For the trained evaluator, this free-flowing speech of level six is easy to recognize. It is not confused with a deviant speech pattern.

The best place to observe level six people is by watching the Olympics. The athletes are coordinated; their bodies are like a miracle of symmetry—working the right way. Their speech patterns fit into this overall picture. Think of how they win by 1/100 of a second. This could not be done with something wrong in the body; it has to be coordinated and functioning the right way. An uncorrected or covered up speech problem could keep an athlete from competing in the Olympics.

Level five is next on my diagnostic scale. Individuals in this group have some deviation of their articulation and speech flow. They represent about 20 percent of the school population. They would benefit from some help, but most professionals and parents would question me on this. These children appear to be doing their best.

Level four makes up about 40 percent of the school population. Their deviations are noticeable to the average observer. They have minor problems with reading, writing, and other school subjects.

The remaining students—about 39 percent—make up levels one, two, and three. They have very noticeable language and articulation difficulties that are detrimental to their education. But our present-day education system only allows the bottom one percent to receive help. The right help for the majority of students would bring about the success our society desires in educating students. It also would greatly lower our drop-out rate and the amount of frustration we see in students today. Today most of our education funding is being spent on the bottom one percent. We would get more for our money by helping the middle 40 percent. We waste our money by using the current remedial methods for children who do not have the intellectual ability to compensate for their problems. These children need their problems cured, not covered up and compensated for, if they are ever to meet their potential.

Our schools try to solve educational problems with better teachers, more money, more administration, and awards and bonuses for better test scores. We cannot get the students in levels one through four to their potential unless we become cure-oriented, however. Until parents and politicians focus on cure, the level five and six students will rule.

The practice of grouping children by their symptoms needs to stop. Symptom-treating with therapy and remedial classes may teach children to compensate for their problems, but it does not help individuals meet their potential. They are not cured, and most are unable to proceed without special attention and after much time and money is spent on them. My theory is not adaptable to group therapy, because no two children are alike. No two children are corrected in the same way. Although the causes of their problems are similar, individuals are unique in that events happen differently during their developmental stages.

Gaining a New Freedom

What I say about the education of children applies also to adults. Heidi, the college junior, was still fighting her old speech pattern. The casual observer could not detect her struggle. On the surface she seemed to be speaking well, but her speech had not been corrected internally. Heidi would have been rated on the lower half of my scale in kindergarten. Had my therapy been available to her then, she never would have developed the severe comprehension problems that distressed her throughout school. She would not have needed to hesitate in reading or work to form her words. Her reading would have flowed. But Heidi had to work extra hard to say words acceptably. In fact, she could pronounce some words more easily than I could. But I could comprehend them; she couldn't.

Heidi had to return to her developmental period, when oral habits of sucking on her index finger and biting her fingernails caused her to misarticulate the *t, d, n, l,* and *r* sounds and to shift her jaw on the *s* and *z* sounds, causing them to be distorted. After reestablishing her articulation skills and reconditioning her reading skills, Heidi became fluid when she read, and comprehension followed.

Heidi passed her political science course, then visited me to let me know how great things were going for her. She assured me that she had kept her fingers away from her mouth, then surprised me by saying that she was enjoying her new freedom. I asked, "What freedom?"

"A friend sent me two new books for Christmas, and I haven't read them yet," Heidi said. She said this like she had really conquered something. As a former compulsive reader, Heidi would have had to read those books at least once as soon as she got them. I asked, "Why haven't you read them?"

Heidi said, "I have so many other things to do." With a little smirk and hands on her hips, she added, "I'll read them when I want to."

Improving the Voice

Another example of an adult who struggled with her "corrected speech," then used my therapy program to establish the pattern she was destined to have, is the Japanese mother I presented earlier. Remember how she came to me after realizing that her two sons spoke without effort, but she still struggled after having had her speech corrected many years earlier? This woman was able to do therapy on herself with little instruction from me. I simply told her to practice speech using her thumb as her children did. When she did the therapy while singing, it greatly improved her voice too. I advised her to watch for any remaining habits that might interfere with her progress or prevent her from conditioning the sounds into her system. The habits that produced the errors in her sound system could still be active. Only two months later she returned to my office. I thought she wanted more help on her speech, but I saw that it was coming in beautifully. What she wanted was help with her junior college singing course. I told her to sing with her thumb in her mouth until her voice sounded better with the thumb in than with it out. Then she was to continue therapy until she could produce the improved voice without her thumb.

The next time I saw the Japanese mother, she had not only passed her class but carried out the therapy until she could project well without her thumb. She was delighted to be able to speak and sing without effort.

Heidi referred a voice student from her college. The student was a 1-2-3 Bach singer. Her voice was not as full as it needed to be and kept her from getting a much-desired lead part. She had a beautiful voice, but her professors said she needed more "ump" to it. I had her practice once a week for three months under my supervision. She acquired the projection she needed and maintained the feeling that produced the quality sound that came out of her. The student was very pleased to reach her optimum performance level and secure the lead she so much wanted.

The majority of voice problems are no more than habit-induced problems. An example is a lady who called me on the telephone for an appointment. She spoke in such a rough, brassy voice that I called her "sir."

"I'm a she," she replied.

I apologized profusely, but she assured me that others made the same mistake. She said, "If I do not change my voice, I am going to lose my job."

"What is your occupation?" I asked.

"I'm in charge of a fat farm. Actually I'm an instructor for overweight people, and I have to holler to get them to do what they are supposed to do," she said.

During our third visit, I asked about her real occupation. "I paint. I am an artist. I do mostly flowers and scenery. My job makes it possible for me to afford painting," she said.

I told the painter that her voice delivery was very similar to a person who talks with a pen or pencil in the mouth. She looked at me, then lifted an imaginary paintbrush to her mouth. "This is my third hand," she said, referring to her mouth. "I can't use my third hand anymore, can I?"

About three months later the woman I had called "sir" could talk like a lady. By stopping her habit, she easily reconditioned her voice for the quality she was designed to produce.

Find the Cause

Whether a person's main complaint is physical as in singing or speaking or mental as in processing information, it is crucial to find the habit or habits that cause the problem.

The causative habits that I refer to are not simple repetitive behaviors that many people think of when habits are mentioned. They are consistent and persistent behaviors that require the body to change the way it operates. They prevent the body from functioning according to its intended design.

How habits affect mental functioning is not so obvious, but an example may help clarify how a learning problem develops out of a speech problem when the causative habit is not detected and eliminated: Consider a child who says "wun, wun wittle wabbit" in place of "run, run little rabbit." At some point in his development, someone says to him, "No, that is wrong. It is 'rrrun, run little rabbit.' Say it the right way for me."

He'll likely answer, "That's what I said. Wun, wun wittle wabbit!" It may take months to convince him that his speech is wrong. When someone teaches him to make an *r* sound, it feels funny to him. It feels foreign and difficult. The child is almost afraid to use the new production, because he thinks other people will think that he sounds funny—not that "wun, wun, wun" sounds funny but that "run, run, run" sounds funny. He has to be thoroughly convinced that "run" is correct. Eventually he accepts this and controls his own comfortable production of "wun."

Another example is a child named Billy. The principal called me to his school because Billy was being teased daily for his speech. Billy would say, "Biwwie wikes yemon yewow" for "Billy likes lemon jello."

Older boys had Billy stand against a wall on the playground and repeat sentences after them, then they laughed at him. Billy mistook this attention as being special. The principal took me out on the playground to watch one of these encounters. "I want this child corrected," he said. "I understand you are doing something special with children. My therapists are not able to help him. Please work with him."

I discovered that Billy kept a knuckle in his mouth, causing his tongue to flatten. I convinced him that the knuckle habit needed to stop. Then I worked on reconditioning his speech pattern. He was talking very well when the principal called me again. We went onto the playground to see Billy standing alone. The principal could not make sense out of it. "Look, Billy comes out for recess, goes over to the wall, and just stands there."

I thought, *Billy is standing there waiting.* I went over to talk with him. "What's the matter, Billy?" I asked.

"Nobody comes anymore," he said. "How come?"

I had to tell Billy that the children had been making fun of him. Now that his speech was corrected, they did not have anything to laugh at. I suggested to Billy that he join his classmates for playtime. Billy walked right over to his friends and joined in. His speech was corrected, and his habits were no longer detrimental to him.

The Roger Family

Some children, even adults, appear to have speech problems but actually imitate another's speech pattern. The Roger family is an example. Five of their seven children misarticulated the *t, d, n, l,* and *r* sounds and distorted *sh, zh, ch, j, s,* and *z* sounds. They could not pronounce their own family name. They all sounded alike.

The parents wanted help for their children but could not afford to have five children in therapy, so I visited only one time with the entire clan. I looked at their speech and habit patterns. For three of the children, the habit pattern did not match the speech pattern, so I knew that they were imitating someone else's speech. With this particular family, the oldest child and the middle child were habit locked. The other three were imitators. I told the parents to have them knock it off. The two with corresponding habits needed the treatment to stop their habits and to recondition their speech.

Under the old speech-therapy system, the one still used today, there is no way to know who needs therapy and who doesn't. All five children mentioned above would be enrolled in therapy. The oldest child with the undetected causative habit would soon be dismissed as being too old to correct. All the children would be tested for language development, and very likely the two youngest would get therapy.

Symptom management for speech and learning problems only covers up the cause. It does not cure. Cause-oriented diagnosing is important. It separates organic conditions from the nonorganic conditions such as learning and emotional problems.

The pebble in education? It could be a sock. It could be little fingers or toes. Consider what goes on in the classroom, that is, what goes into the students' mouths: pencils, clothing, paper clips, jewelry, purse strings, paper, hair clips, combs, shoe strings, erasers, fingers, fingernails, knuckles, hair—the list is endless.

These little things, like the pebble in a shoe, can make basic functioning difficult.

Part Three

THERAPY

Chapter Eight

The Thumb and Therapy

Vaccine therapy. The big break that I prayed for, the conditioning agent that I knew I needed to find, was put before me as a vaccination. This big break that the cure would be a vaccination in the form of a thumb, as I mentioned in chapter one, came as I focused on learning the cause of speech disorders.

Not only had I discovered that oral habits cause speech problems, I found that the very causes could be used as the cure. After I applied my vaccination-type therapy to speech cases, all the other benefits came to light. Numerous other habit-caused symptoms were being cured along with speech symptoms. Vaccine therapy enables individuals to function according to their own correct biological patterns.

How does it work?

I begin by having a student lie down on his back on the floor. I then ask, "If you were to suck your thumb, which one would it be?" By watching body language, it is easy to establish preference immediately. But I ask the student to put the opposite thumb in his mouth.

I observe the body's reaction. It shows me how the muscles are conditioned, and it reveals abnormal patterns that confirm conditions such as the tooth wear, facial grimaces, jaw shifts, tongue retraction, and tongue sidedness (asymmetrical function). In other words, the body's responses reveal the neurological involvement.

How the thumb is placed in the mouth is crucial. Without guidance, students come up with numerous variations. Only a trained therapist can assure that the thumb is positioned correctly, without being turned, without pressing down on the tongue, for example.

I ask the student to close his teeth on the thumb. As the thumb is removed, I have the student close his lips, maintain the jaw and tongue position that the thumb established, and swallow three times. The swallows are not forced. We wait for the natural occurrence. In the meantime, I check the thumb for bite marks—evidence of jaw tension. Some people clamp the thumb so tightly that they draw blood.

Any swallowing difficulties become visible during this first therapy session by using the unpreferred thumb to expose deviant patterns. The preferred thumb is then used to establish correct biological patterns for that individual. The preferred thumb is used to condition the mouth for the correct bite, chewing motion, and swallow pattern. And it is used to train the reflexes for the correct speech pattern. I have students produce sound combinations, then flowing speech with a thumb.

All the cases I have described in this book were treated in the same way. Therapy affects each individual differently, however, because each person has deviated from correct functioning by his own habit patterns.

The therapy is designed to first eliminate the misguiding habits and tension, then to redirect the body to function according to the way God designed it. Although correct eating and speaking are the same for everyone, the outcome certainly does not produce identical creatures. The therapy puts each person in harmony with his own neurological patterns—his own voice, his own unique characteristics. It restores what the habits wrenched from him.

Though the steps of therapy are the same no matter what symptoms are presented, the changes that occur as a result are different for each individual. But everyone who follows the therapy is cured of habit-caused symptoms.

It may seem that I should present all the steps of my therapy so that this chapter could serve as a self-help program, but unfortunately we cannot heal ourselves. We cannot even diagnose our own habits. No one cured by this therapy knew about his particular habit before therapy. No one with multiple habits knew the sequence of a developed pattern.

For example, Sally feared opening her jaws. She panicked one time when she saw a billboard display of a Big Mac. Sally ate by pushing pea-sized bites of food in between her teeth, and she used a straw for drinking. Because of migraine-like headaches that made her physically ill at exam time, Sally was flunking out of school. She was a very uptight college sophomore. Why?

Sally habitually locked her jaw shut with so much force that her jaw muscles had become hardened like iron. She also held her shoulders in an uplifted and forward position with her neck forced back. She used this habit pattern when studying for tests, which resulted in migraine-like pain, then vomiting at exam time. It took days for the pain to go away and for the jaw to open wide enough for anything larger than a pea to be inserted.

Sally drew blood from her thumb during her first therapy session. It took weeks for her to adapt to a different study habit so that her thumb could be easily inserted into her mouth to recondition her system. In fact, she found that other activities and emotions triggered the habit pattern. Happiness and sorrow could trigger the same muscle responses. The response pattern ruined dates and other social events for Sally. She helped me see why tension patterns would return in some people who had the most obvious situations under control.

I taught Sally to bite, chew, and swallow food correctly and to relax her muscles. When she achieved the relaxed mode, her migraine-like headaches disappeared. Then one night, to her roommate's surprise, Sally yawned. "You yawned! You opened your mouth!" the roommate exclaimed. In nearly two years of college, the roommate had not seen Sally's mouth open. Sally had changed.

I used the same bite-chew-swallow therapy on Linda, who had complained that her jaws popped so loudly when eating that she was embarrassed to eat in a restaurant. Linda's brother, an osteopath, could not tell what caused the popping. If the problem wasn't alleviated, he worried that bone damage would result.

The habit was not easily determined, but the clues to its existence were obvious. Linda's tooth wear suggested her jaws came together in an odd position. During our initial visit, Linda could not move her lower jaw into a position that matched the wear pattern. "What do I do now?" she asked.

"Watch everything you do," I said. "And have your husband watch you sleep."

It took only two weeks to uncover the situation responsible for the tooth wear. Linda had an arthritic hip. Pain from the hip woke her in the middle of the night. To relieve the pain, she placed a pillow under her stomach and hips. She angled her face into the mattress, with the jaw pushed into a position that accounted for her tooth wear. No other activity produced this jaw position. Yet it really had nothing to do with the hip problem. Linda could stop it.

I taught Linda to relax her jaw, then to relax it open for inserting food, then to chew and swallow without tension. Once she achieved a normal bite pattern, I taught her how to bite off food correctly. Her bite had been thrown off by her habit. Linda had been throwing her jaw out of joint, causing the loud popping and snapping and the resulting abnormal bite position. After she conquered this habit and established the correct eating pattern, Linda said, "It is great to be able to eat out again without bringing attention to myself."

It may seem incredible that applying these bite-chew-swallow steps to temporal mandibular joint problems and pain would work. But they do.

A great percentage of the population does not know how to masticate—how to chew their food properly. It is common to see people biting either from the front or side of their mouths with muscles stressed in some way. Watch people in fast-food restaurants and pizza parlors. See how they attack their food. Observe the minimal amount of chewing that they do. It's watch out stomach, here it comes! Is it any wonder why so many people have indigestion and related stress symptoms?

Al is one of my favorite examples. A stomach sufferer who could devour a Big Mac in two bites, Al had only two teeth that touched together when he bit down. Everything he ate was swallowed in chunks. Al ate toast by folding it twice to form a four-decker bite that he popped into his mouth before taking a drink. Then he'd swallow the toast and liquid without chewing.

I asked Al to cut all his food into small pieces. After he learned to chew properly, he wanted to continue cutting up his food before eating. Al actually learned to enjoy chewing small bites. His stomach problems disappeared, and his teeth came together as a result of therapy. Al needed no medical or dental attention to treat his symptoms. Al was cured by changing his ways.

Years ago a doctor told my wife Rosalie, "You just have to relax." We had five children. He never told Rosalie *how* to relax, nor did he take the time to show her. I don't think he knew how to relax himself. The current methods for relaxation involve drugs or exercises of some sort. I teach people how to relax by not holding their muscles tightly.

Muscle tension is the body's real destroyer. Contrary to popular belief, the tension pattern used does not depend on how good or how bad life is going. I recall one woman who complained of aching all over.

Martha had been to several medical doctors who found no reason for her aching body. When she woke in the morning sore and tired, she felt lousy all day. Martha hardly remembered a day since early childhood that she felt good.

I asked Martha to grab a pillow and lie down on the floor to show me how she went to sleep at night. She curled into a fetal-like position with such rigidity that someone could grab her by the ankle and lift her off the floor. Is it any wonder that pain and stiffness accompanied such an unnatural way of going to sleep? To someone like Martha, a relaxed muscle is an unnatural feeling. Sammy's case explains this well.

I treated Sammy in a school speech program. He had a paralyzing condition of his facial and lip muscles, but it affected only the right side of his lips. It appeared that he'd had a small stroke. In fact, a doctor gave this as a diagnosis and referred him for physical therapy. The diagnosis was wrong. Physical therapy did not help. A habit had caused Sammy's paralyzed-like lip condition.

To understand how a habit can cause such conditions, Sammy's habit can be imitated to temporarily produce the same symptom. Here is a harmless experiment: Squeeze one side of your upper and lower lips together using the thumb and index finger of one hand. Press hard and long enough for the lips to go numb. Next, stand in front of a mirror, continue to hold the lips tightly, and talk for awhile. Then let go of the lips and continue talking. The resulting pattern you see is Sammy's presentational speech production.

I discovered Sammy's habit by observing him in the classroom. He held the right side of his lips together for 20–30 seconds at a time. I mentioned this to the teacher. She said, "Oh, yes, he does that all the time, especially when reading—even when reading out loud." Later I learned that Sammy went to sleep at night pinching his lips.

I worked with Sammy, his teachers, and parents to help him stop the habit. One day during a therapy session, Sammy walked over to a mirror and flipped his lips with a finger. He said, "They feel like two loose pieces of meat, but they don't look that way." He was right.

Muscle coming out of tension does not feel normal. The natural feeling has to be incorporated both physically and psychologically. By use of the thumb, Sammy reestablished a natural muscle function; he retrained his lips to function normally. No longer did Sammy have one-sided speech that he needed to explain as "probably being caused by a stroke." I feel that Sammy's medical diagnosis should have read: "Cause of problem unknown." I wonder why it was necessary to assume a stroke.

This therapy is so unique that I have to sometimes stand back to see the trees—to see how trees make a forest—to view the overall picture. I sometimes have to do this in therapy, that is, go back to when a student's progress stops. One of two things can be happening.

First, the student may not be practicing for any number of reasons. If so, it is obvious. There is no change. This is a benefit to the therapist who can confront the student with either, "You are doing a nice job" or "Why are we wasting our time?"

Second, it is possible that something has been overlooked. This involves the sequence of the habit patterning—when another habit that came before the more obvious one has not been exposed. Secondary habits show up in the therapy process. Usually there are no more than two conditioning habits per condition.

This therapy has two other unique factors. First, each therapy session meets the student anew. The current status of the person is assessed before moving on. Second, any fully-trained therapist can continue the same therapy without any past information. Records only need to be kept for administrative purposes. When therapy is concluded, it is as though no problem existed. The body is in its natural mode. With present-day therapies, this is not true.

It is possible to do a seventh- and eighth-grade screening in a school and pick out the students who had been provided conventional speech therapy and dismissed as corrected. If the school has more than one speech therapist, it is possible to go a step further and pick out which students had which therapist. This is done by observing the speech patterns of the therapists and the patterns of the corrected students.

For example, if a therapist makes the s-sound with her front teeth coming together, her students use the same position to produce the sound. Or it may be a therapist's jaw shift, some type of sidedness, or a facial grimace that could be incorporated into the students' productions. This is the result of the speech-imitation approach—the visualized-therapy approach—which is a superimposed production rather than a natural production. A copied pattern never becomes natural for a student.

The vaccine, habit-treating therapy allows for natural physiological patterning to occur. This cannot happen by introducing extraneous patterns that interfere with the natural sequence of events. Rather, patterns superimposed onto an incorrect internal reflex pattern create a handicap on top of a handicap.

If the student merely imitates someone's incorrect speech pattern, that is, his production is not physically induced by a habit, then his correct pattern is restored by merely getting him to stop imitating the defective pattern. No therapy is necessary. His internal pattern is not disordered.

Everything I have discovered is contrary to my training as a speech pathologist. I was told that there is no universal therapy or cure. During a conference, a Ph.D. in the speech field told me that the cause of an individual's speech problem happens long before treatment begins and does not matter. The professor completely cut me off. I don't think he even liked his answer given that he didn't know the causes, so the discussion ended.

I found the answers myself. The cure for speech problems is universal. The causes do matter. The causes don't go away and leave speech disorders behind. The causes, in fact, must be treated to obtain a cure.

As with speech therapy methods, my vaccine therapy differs from current methods used in diagnosing and treating habit-caused dental disorders. Incorrect muscle patterns created by oral habits cause overbites, collapsed arches, tooth crowding, and other dental conditions.

I have done this therapy for 40 years and estimate that 85 percent of four-bicuspid extractions result from nail-biting habits. When the orthodontic treatment is finished, removal of impacted wisdom teeth soon follows. This maiming process produces an old-age appearance years before the patient's time. It also has produced a joke-like comment heard in dental, especially orthodontic, offices: "Four on the floor and a nose job"

The "four on the floor" are the four removed bicuspids. The "nose job" refers to the need to treat a resulting bump on the nose. The age when the teeth are usually pulled causes an abnormal reduction in the growth patterns of the oral structure. This leads to the skeletal feature of the bump on the nose. It occurs where the bony ridge of the nose ends and cartilage begins. The orthodontic procedure that pulls the mouth back creates the bump. Plastic surgery is a likely step to follow.

The pulling of healthy teeth and the banding and bracing of remaining teeth result in impacted wisdom teeth. The result? Additional shrinkage.

The removal of four wisdom teeth and four bicuspids results in extreme shrinkage of the mouth. To imagine these eight teeth in the palm of a hand illustrates how much corresponding bone loss occurs in the dental arches. This leads to quite a deficiency in the oral structure. Recall the British study of the twin girls. The one who had teeth removed grew ugly while her sister, who was treated for the same dental condition without the removal of teeth, grew pretty.

In the next chapter I address therapy and dental conditions in more detail. I present pictures of habit-caused conditions, and I present progress pictures showing changes in the dentition. The trained therapist learns to recognize physical clues in the oral structure to pinpoint the habits that produce abnormal conditions.

The therapist does not cure anyone but rather guides students along the healing journey. He knows that emotions are often involved and prepares students for emotional release. He knows how to use the thumb to prevent and to cure deviant functioning patterns. The therapist guides students in using the thumb to correct speech patterns, even for reading and learning. Once the sound system is corrected, a student is taken through reading levels—from a primary reading level to his current level—to reintegrate his system for correct neurological patterning. All this is done with a thumb.

Who would have thought that a thumb could change a person's life to such a degree? There are many adult thumb suckers today, many more than we can imagine, who still use their thumbs to go to sleep at night or to achieve a degree of good feeling when stressed. These people have found a way to meet daily stress, but they know that something is missing. They continue to use the thumb to maintain a balanced existence, and most feel guilty about it.

This therapy sets the stage for alleviating the need for thumb sucking and all other oral habits by correcting imbalances. The person who gives up an oral habit through willpower may do himself an injustice—another concept that contradicts current thinking!

Even habits such as smoking and drinking given up by willpower rob the person of ever knowing what relaxed living is all about. The vaccine-type therapy removes the need for a habit and relaxes and restores harmony to the body.

Our bodies are an amazing creation. When guided away from misuse and helped to establish correct use, a natural restoration unfolds. Creation's healing power is a marvel to behold.

Chapter Nine

Dentistry and Therapy

During weekdays, Dr. Berke and I rarely had a chance to visit. We were both very busy, so we scheduled Saturday afternoon appointments for cases that we wanted to study together. Before we began seeing the patients, we purchased our favorite snack—peanut brittle—something we never recommended to our patients. We'd enjoy chomping on the peanut brittle while we prepared for the afternoon.

Dr. Berke, who is a children's dentist trained in orthodontics and my mentor in the field of dentistry, reviewed each case with me. Together we'd examine the cases. Often he asked about the patient's habits and progress. He marveled at the progress that I achieved by using the same therapy on all types of cases. Seeing how quickly the corrections took place also surprised him. He noted tooth movement. He said that if he moved teeth that quickly with orthodontic treatment, he would cause great injury to the teeth.

We used the therapy for all types of orthodontic problems—crossbites (right side, left side, or both sides in cross), open bites (anterior, lateral, or bilateral), deep bites, overjet, and under bites, and any combinations of these bites. Many individuals have open and crossbite combinations as well as other combinations. Just use your imagination.

Most orthodontists do not even pretend to be able to correct open and crossbites with their treatments. Yet the therapy corrected the bites and maintained the corrections.

The reason the same therapy can be used for all of these conditions is that the muscle patterns and outside forces involved with oral habits are responsible. Very seldom do anatomical defects cause these conditions. When my therapy is followed, the corrections occur at a faster rate than they do with treatment using bands and braces, and there is no damage caused to the structure. The younger the child begins therapy, the better the result. In fact, this therapy can be initiated at birth; orthodontic treatment should not be considered before full growth, about age eighteen.

The cessation of the causative habits usually produces a correct dental occlusion that will service any individual for the rest of his life, barring injury or disease. My therapy, when properly used and carried out, gives an individual the occlusion that nature intended him to have. The correction is permanent unless the individual acquires new detrimental habits or returns to old habits.

As long as an individual continues to operate as trained in therapy and does not adopt another habit, his corrected condition remains. I'll give two examples.

Peter, a ten-year old, successfully went through my program and maintained a healthy bite for three years. Then his mother brought him in because an upper tooth had started to protrude and a tooth below it tilted inward. Obviously a new habit, involving a prying force, had caused this. I suspected anything that fit between the upper and lower teeth. Since only two teeth were involved, it had to be an object small enough to fit between the two teeth—something like a pencil or little finger.

Peter and his mother spent half an hour trying to figure out what the object might be with no success. As they were getting ready to leave and the mother's hand grasped the doorknob, Peter and his mother both turned around and said simultaneously, "The Popsicle sticks!"

I asked, "What do you do with Popsicle sticks?"

Peter described his new ritual of placing a stick between his teeth, then pulling downward to break it. He broke the sticks one at a time into one-inch pieces. The habit had moved the teeth within three months.

Peter did not have to give up his daily treat, just the habit of breaking the sticks with his teeth. Within three weeks his teeth shifted back into place.

Another patient also returned for an additional therapy session. Martha, age thirty-five, had successfully corrected bilateral open bites that were the result of an occlusal adjustment. Her teeth had been filed down and a palate spreader had been applied in an attempt to improve her bite. But her teeth had spread outward too far. I had her get rid of the palate spreader. Through therapy she was able to close the bite. A year later Martha returned because her two upper lateral front incisors started to protrude forward from the arch line.

I discovered that Martha had been promoted in her job at a department store. She explained that she was now responsible for several stores, requiring her to unpack clothes and hang them on racks. I asked her how she went about the task. Martha grabbed an imaginary hanger and started to bring it to her mouth. Suddenly she realized that using her mouth as a third hand had caused her problem. This new persistent habit had altered her teeth alignment. When she stopped the habit, her teeth returned to their correct position.

It is amazing to see how any habit-caused deviation spontaneously corrects without any assistance other than the natural muscle forces in the oral structure. What orthodontists usually see in tooth positioning and occlusion problems, Mother Nature can correct without bands and braces. With all unnatural forces removed (habits or even braces), our bodies correct even the most serious of dental malocclusions because it takes the oral habits to maintain malocclusions. Once detrimental habits are eliminated, the body can heal itself.

Oral Habits and Growth Patterns

Let's look at an artist's illustrations of what oral habits could do to one person's facial structure. Imagine five identical quintuplets, each developing different oral habits that affect their growth potential.

Quint One grows up with no detrimental habits that disrupt the intended design of her oral structure. No speech problems develop. She meets her maximum potential. She becomes a beautiful and talented young woman.

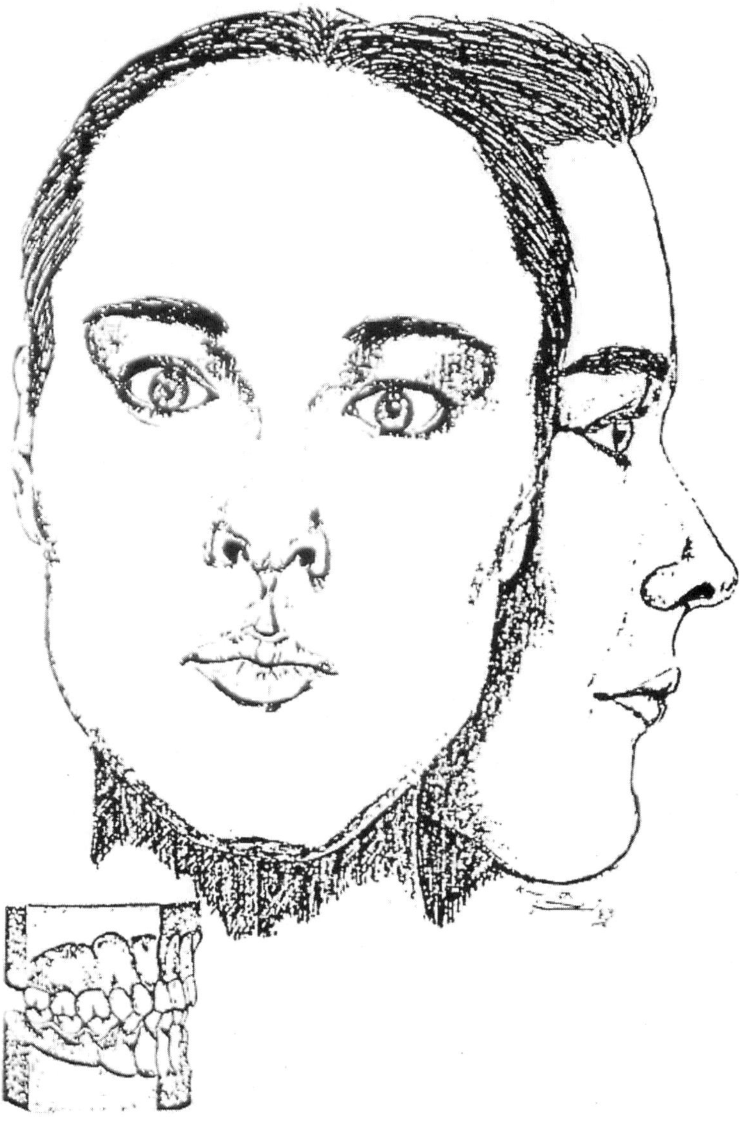

As a child, Quint Two sucked two fingers in a manner that caused a protrusion of the upper anterior teeth and a bilateral crossbite, which made it next to impossible for her to bite and chew her food properly. The improper chewing pattern then created further damage. She also developed speech and learning problems. She became very jealous of her sister. As you can see from the sketch, she did not grow as lovely as Quint One did.

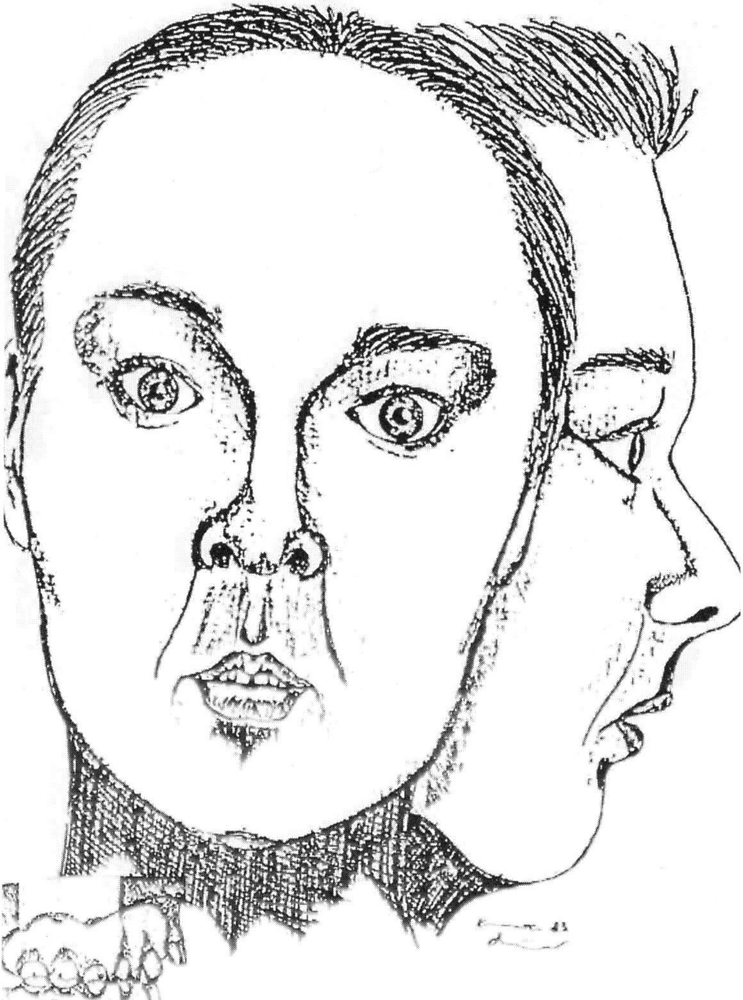

Quint Three placed her index finger in a palm-up position against the palate and sucked it. The weight of her arm added enough force to the inserted finger that it directed the palate's bone growth up and forward, affecting the nasal passageways for air flow. She became a mouth-breather.

The mouth breathing that the habit caused was detrimental to biting, chewing, and swallowing. This also produced problems with speech, which, in turn, caused learning and reading problems. The destructive habit also caused the upper arch to collapse, leading to a severe bilateral crossbite that also affected her appearance. This quint became very self-conscious, somewhat antisocial, and jealous of both Quints One and Two.

Quint Four learned to imitate her mother's "really get mad" look in which she threw her lower jaw forward and grimaced. Since she had four other sisters to contend with, she was mad all the time. This habit caused her lower arch to capture the upper arch. Her upper wisdom teeth had no opposing molars from the lower arch. This caused the upper wisdom teeth to lock in behind the lower wisdom teeth. Surgery would be needed to correct this condition. Fortunately she had no speech or learning problems, although her speech delivery possessed a different quality. She was not jealous of her sisters. She felt sympathy for Sisters Two and Three. She was concerned for herself regarding the idea of surgery.

Quint Five developed habits that added a lot of tension to her mouth, leading to tooth crowding. She wanted to look like Sister One. In fact, with her mouth closed she did look like Sister One. Because she wanted straight teeth, she pestered her parents for orthodontic treatment. The orthodontist removed four bicuspids in order to realign her teeth. The straightening of her teeth led to the need to remove her wisdom teeth later on, however. This caused extreme shrinkage of her dental arches, reducing her mouth size to that of an eleven-year old. The reduction of bone support aged her adult face. Her profile collapsed because an adult dental structure no longer supported her facial muscles. This caused her nose to appear to protrude forward and to develop a bump. She looked about six years older than Sister One. She even developed an aversion to being seen with any of her sisters because people assumed she was older than her identical sisters.

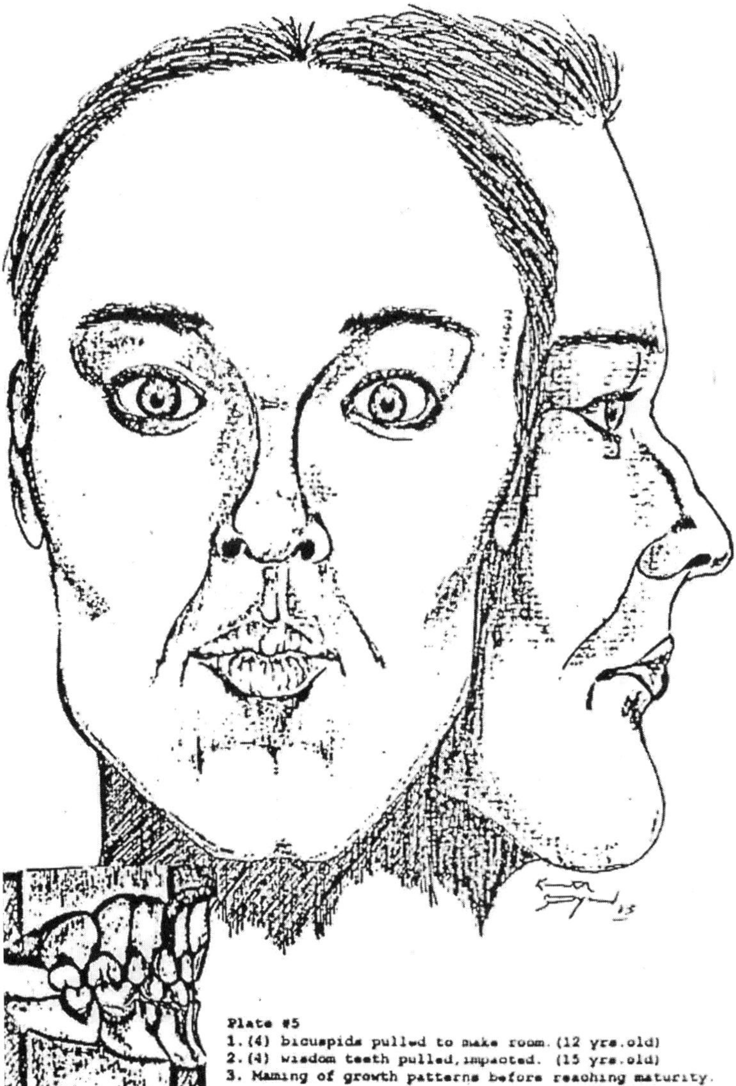

Plate #5
1. (4) bicuspids pulled to make room. (12 yrs.old)
2. (4) wisdom teeth pulled, impacted. (15 yrs.old)
3. Maming of growth patterns before reaching maturity.

The quintuplets, all five of them, could have remained identical in appearance throughout their lives had not four of them developed individualized oral habits that misdirected their growth and development. The artist's drawings showed us, through Quint One, the potential we have if we stay on the correct path. Through the other four quints, we see what can happen when we deviate because of injurious habits.

Look again at the sketches of the teeth. See how Quint One's full complement of teeth supports a full facial structure. For the other four quints notice how their altered facial structures conform to the position of their teeth.

It is sad for me to think that millions of individuals suffer from similar conditions that could be corrected through my therapy—beautifully corrected if seen in the formative years. Actually all such detrimental habits can be prevented if parents learn how to catch and stop them. This means that parents could prevent damage to the oral structure and prevent speech and learning problems from developing in their children. This would, in turn, benefit their emotional health. One of my goals is to teach parents and other educators and health-care providers how to prevent such problems.

Four Case Studies

The following pictures are cases of young children, ages 4–8, who were corrected by my therapy. Orthodontic treatment was not needed.

CASE 1

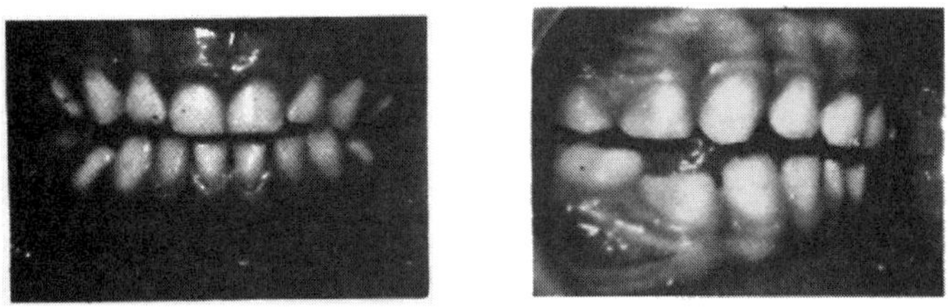

Before Treatment:
- Anterior and posterior open bite
- Ankyloid first molar
- Lateral crossbite
- Age 5 1/2

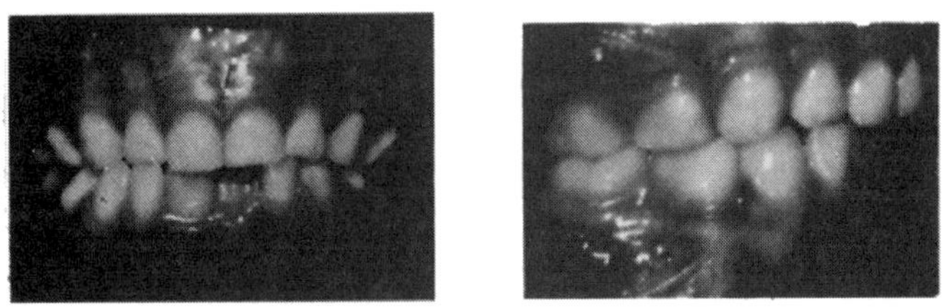

After Treatment:
- "Ankyloid" first molar errupted
- Anterior and posterior open bite closed
- Treatment time: 11 months

CASE ONE

This child's teeth were in a crossbite, and he was growing his face crooked. One baby molar had failed to fully erupt. The top pictures show the relationship of the teeth when the jaw was brought to center position where it should function. The pictures reveal that the actual condition is an anterior and posterior open bite.

This child's crossbite condition was an example of an open bite gone bust. That is, in order to bite off and chew his food, he was forced to throw his jaw into a crossbite to bring his teeth close enough together for eating. In the cross position, his eye tooth locked down on the molar, keeping it from erupting. Such a tooth is often misdiagnosed as an ankyloid tooth, one that has attached itself to the bone. Because the child used his facial muscles with the jaw pulled to one side, his face was growing crooked. He also spoke with a frontal lisp.

The child started therapy at age five and completed it eleven months later. I taught the child to bite and chew his food with his jaw on center. Because of the open bite, he could not handle certain foods. We modified his diet until his teeth closed, making it possible for him to bite and chew correctly. Many foods initially could not be chewed properly, and they had to be cut into small pieces to help speed up the process of closing the open bites.

This child caused his dental and speech problems with a nail-biting habit and further hindered his bite with his eating habits. He first created the anterior open bite and the frontal lisp by using the habit at the front of his mouth. As the bite opened and he no longer could bite his nails in front, he began biting them on the right side, thus bringing his eye tooth in contact with his first baby molar. This led to his moving the jaw to the left for chewing, thus collapsing the bite in the cross position. He could not speak from this position because his speech had been learned from the open position. His jaw went back to center for speaking but otherwise operated to the left.

Through therapy, the child learned to use his facial muscles for speaking and eating in the correct way. The lower set of pictures shows how his bite closed within eleven months of therapy.

CASE 2

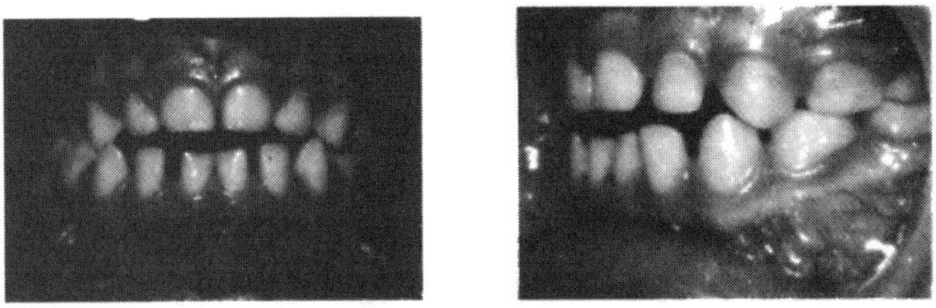

Before Treatment:
- Open bite
- Age 4

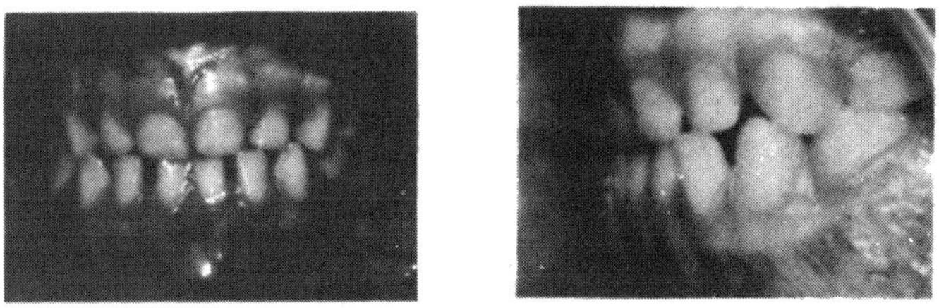

After Treatment:
- Open bite closed
- Treatment time: 6 1/2 months
- Note: Correction is maintaining now at age 7

CASE TWO

The top pictures show the results of a child's habits of sucking on a knuckle and biting on fingernails. The nail-biting habit threw her into a slight crossbite. Both the nail and knuckle habits contributed to the open bite. The child was first seen at age four. She had a frontal lisp. It took just six months to correct her speech and the dental bite as seen in the lower set of pictures.

CASE 3

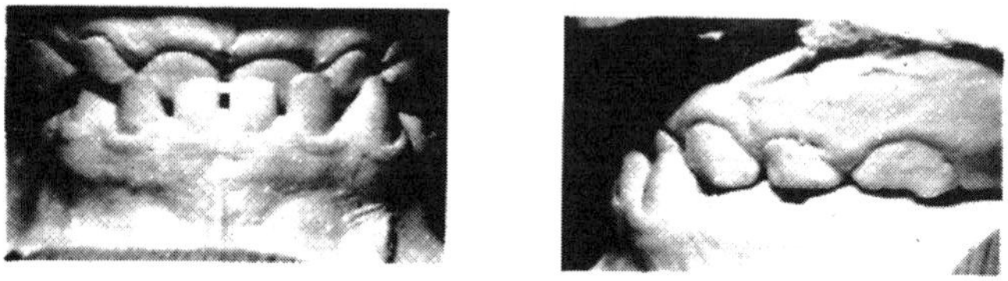

Before Treatment:
- Skeletal Class 3
- Age 4

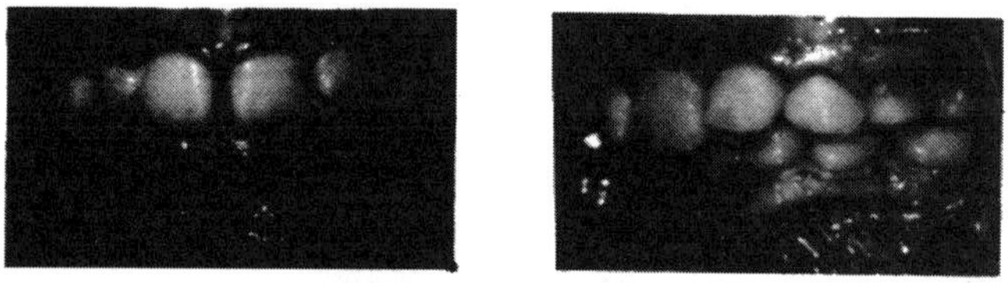

After Treatment:
- Skeletal Class 1
- Treatment time: One year
- Maintaining correction four years later

CASE THREE

The top pictures were taken of the original dental molds on the little four-year-old girl—the "bulldog jaw" case—presented in chapter five. She was given a dental diagnosis of Skeletal Class III after panoramic pictures and measurements were taken and interpreted according to current orthodontic practices.

This child stopped her habit of throwing her jaw forward to allow her body to correct itself. Dr. Berke and I observed the correction in progress. We watched as the upper alveolar ridge became freed from captivity by the lower jaw. The upper anterior teeth then raised up and went over the bottom teeth. Natural skeletal growth aided the process. She obtained this correction within one year. The lower pictures, taken four years later, show how the correction remained. The diagnosis of a Skeletal Class III growth pattern was no longer valid, and the child did not have to face surgery.

CASE 4

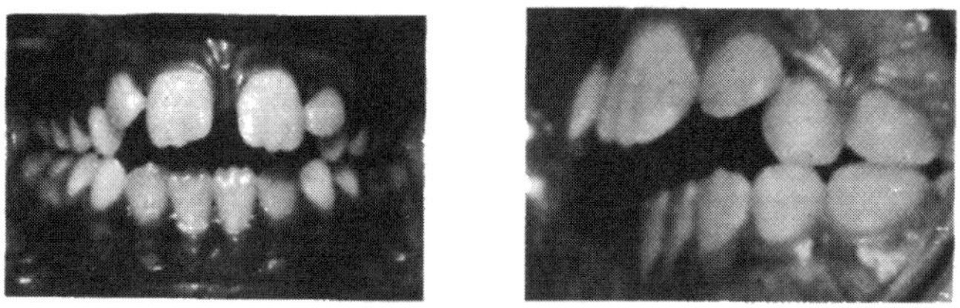

Before Treatment:
- Anterior open bite
- Age 8

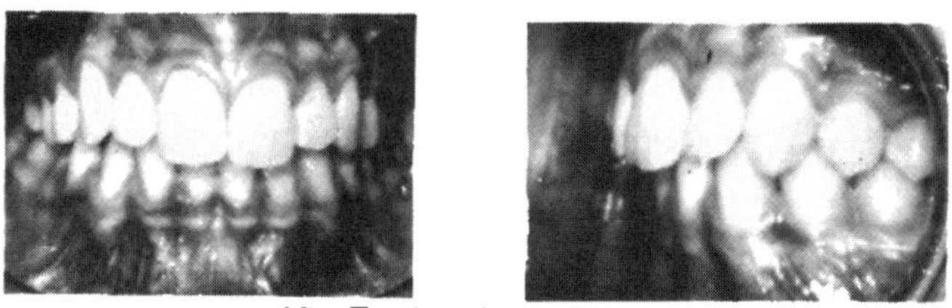

After Treatment:
- Open bite closed
- Treatment time: 7 months
- Maintaining correction 4 years later

CASE FOUR

The top pictures show the open bite of an eight-year-old girl, which was caused by sucking on her second and third fingers. Because she sucked the fingers far back into her mouth, she incorrectly produced many speech sounds. It took just seven months for this young girl to close her bite and correct her speech pattern. The lower pictures show how she maintained her corrected position after four years.

These four cases demonstrate the power of the therapy to correct and maintain a natural, correct bite for the rest of a person's life—if there is no return to old habits and no acquiring of new habits injurious to dental health. The bite, chew, and swallow regiment of this therapy is integrated into a life-supporting way of life. The individuals are not controlling the way they function; they are operating according to the way they are designed to function.

These cases were, at the time, considered to be feathers in my cap, not only for the corrections but for the short period of time it took for these individuals to accomplish the corrections. In fact, what the child with the bulldog appearance accomplished has been considered impossible by orthodontists.

It was not until many years later that I realized my therapy—therapy that will correct the almost impossible—was not good enough. I discovered that it did only half the job.

Within the past few years I realized how detrimental habits can cause learning problems, and how I could correct those problems. During my forty years of clinical practice, there have been many individuals whom I treated only to the point of correcting their dental bite and speech, but whom I did not assist with learning difficulties that likely existed.

I no longer consider therapy to correct just dental and speech problems to be sufficient to help produce students who can reach their maximum potential. The therapy must include the reconditioning of the system for reading to truly help students meet life's challenges with their true potential.

A Plea to Parents

Your children could very well hear comments from their peers, such as "What's the matter? Can't your parents afford to have your teeth straightened?" This may lead you to rush to an orthodontic office where you will hear how bands and braces can straighten your child's teeth and correct most other dental problems.

The bill will be $3,000 or more, depending upon your child's cooperation with the orthodontist. If you go ahead with orthodontic treatment, I advise that you get a loan to pay the orthodontist up front. This keeps your child from being held in bondage longer than necessary; it assures the most prompt treatment and less time in bands and braces. This, in turn, gives the least amount of damage to the immature enamel on your child's teeth. If the orthodontic treatment lingers, look for somebody who does the type of therapy presented in this book.

Now ask yourself the question: Why doesn't the orthodontist learn methods that correct bites without using bands and braces? Before you consider allowing any teeth to be pulled, look again at the last profile of the drawings I've included in this chapter. Do you want to maim your child by having teeth extracted? Do you want to allow facial deficits that will make your child look years older as he or she matures? Look at the bite in the drawing to see just how much tissue is lost in the oral cavity by the extraction technique.

My sincere advice to you is to find an orthodontist who will not extract teeth but will move them into the proper bite position and provide habit therapy before, during, and after orthodontic treatment so the correction will maintain itself without retainers.

In a future book, I will present a number of cases where orthodontists could not close open bites with bands and braces, retainers, etc. I accepted these cases only for orthodontists willing to remove the arch wires or retainers to allow my therapy to close the bites. Some of these individuals had been in bands and braces for four and five years, a painfully long time.

Please, be warned. Remember that I know about dental conditions from working in dental offices, especially with Dr. Berke, who dared to investigate my work. Dr. Berke taught me everything I needed to know about the dental field and current treatments so that I could consider the problems that orthodontists see in their practices. I had the opportunity to try my therapy on every type of case.

I owe Dr. Berke much gratitude for his role in my discoveries. I know many people who feel the same gratitude, and I hope increasing numbers will recognize his contributions to better dental care in the years to come.

It was with Dr. Berke's support and encouragement that I began offering seminars to dentists in 1982. We called my therapy "myosteodontic therapy" because of its effect on muscle, bone, and teeth, and to distinguish it from myofunctional therapy that focuses on compensatory oral muscle exercises.

I close this chapter with a letter Dr. Berke wrote recommending a seminar to his colleagues. (Dr. Berke has since retired. The office in his letterhead is closed.)

ARTHUR BERKE, D.D.S.
Dentistry for Children
1526 Cravens Avenue, Suite 1
Torrance, California
Telephone: Fairfax 8-3444

September 1982

Dear Colleagues:

John Streicher has been associated with me for the last six years. During that time, he has demonstrated repeatedly that it is possible to correct many mandibular crossbites, tongue thrusts, Class II overjets, nonskeletal open bites, and pseudo Class IIIs, even in *four-* and *five-* year olds without the use of any appliances or mechanical devices.

He has shown, repeatedly, that his myosteodontic therapy not only corrects these malocclusions but also corrects speech problems and myofacial pain disorders (TMJ) by altering oral habits that have been a major factor in their etiology.

These techniques can be used by most dentists in their practice.

I am confident that you will find his seminar most interesting and valuable to your practice.

Sincerely,
Arthur Berke, D.D.S.

Part Four

Some History

Science can change the compound states of matter easier than it can change its mind.

—Calvin Miller

Chapter Ten

The Speech Therapy Field

When I entered his office at the University of Southern California, my counselor was holding the results of my college records exam and wanted to talk to me about my career choice.

"John, what do you read during your leisure time?" he asked.

"I usually choose scientific articles," I said.

"Uh hmm. I would have guessed that from your tests results. You scored extremely high in science. Why aren't you going into medicine? Why are you considering speech pathology?"

He did not consider speech therapy to be scientific. I agreed with him. I said, "I am going to make the field of speech pathology scientific."

The counselor gave me an oh-yeah look but approved my admission into graduate school. That was in 1959.

I had realized the need for making the field more scientific during undergraduate study. After enrolling, I visited therapists in the public schools. All the therapists did a "listen-to-me, look-at-me, now-imitate-me" type of therapy. I already had begun to notice habits, and I particularly noticed the habits of the therapists who unknowingly had children imitate their jaw shifts and tongue movements.

I went on to take classes. Lee Travis, Ph.D., who is considered to be one of the founders of speech pathology and who compiled a thick handbook for speech practitioners, was one of my instructors. I remember exactly where he stood when he lectured. He positioned himself at a window opened to the California smog. One foot rested on the windowsill. One hand embraced a cigarette. I asked my first question.

"Dr. Travis, I have been wondering. Why do some children stick out their tongues on certain sounds?"

"It is, possibly, that these children hate their parents and are just getting even with them," he said.

It wasn't what I expected—not scientific in the least—for such an honest question. A laugh wanted to erupt from inside me, but I had to contain it. The man was dead serious.

I soon realized that USC was heavy into psychology. Speech problems were considered to be psychological in nature. From one professor I learned that anal retention was the most acceptable answer for most questions on every test and assignment. I passed his tests by mentioning it on every exam sheet. What I believed would never have passed. Throughout college I disagreed with what I heard being taught in speech therapy classes, and I disagreed with most of what Travis and his contributing coauthors put in the handbook.

The classes I took as a graduate student weren't any better. Some were in physiology and some in psychology. We listened to quite a few visiting professors. I could not believe the things they came up with. Some talked about brain functions and how speech problems were derivatives of these functions. But they could not handle any of my questions directed at the meat of their theories. They got frustrated with me. They accused me of not being humble enough. I disagreed about that too. I was trying to learn.

I requested to do a study based on my observations. I had the resources and could work a research project into my schedule. I proposed that I do it for my master's thesis. I was told, "No. You have to do what we give you to do."

I was not happy with this decision or anything else going on in speech at USC. I wasn't alone. Eight doctoral candidates were not happy there. Seven transferred to other colleges. I started going to different colleges to hear different professionals in the field. I went to the University of California in Los Angeles, California State in Los Angeles, and California State in Long Beach.

I learned that the doctoral candidates and I were not the only ones to pull out of USC. Las Floristas, USC's funding source, had given an ultimatum because of the number of children failing to improve in the speech program. The USC speech department folded. Then the same thing happened at UCLA. Its speech department could not prove that they could correct speech problems. It lost funding. The department was transferred to California State in Los Angeles.

I took a course from the director of the department at California State. She had been in charge of treating the son of a professor who headed up another department at the college. The little fellow had been in therapy for three years. The unhappy father accompanied his son for a reevaluation. My class watched the evaluation through an observation mirror. I asked if I could meet the child before the evaluation. I was granted a ten-minute visit. I asked the child a few questions and looked into his mouth. His teeth and his speech pattern were almost exactly like another child whom I had seen for therapy. I asked my professor to find out how the child went to sleep at night.

I watched the evaluation with my classmates. The professor followed her evaluation procedure to the letter and never inserted my question. She got up to leave. I whispered under my breath, "Ask, ask, ask." My classmates knew what I wanted and what I expected to learn. They were on edge with me. Then, as the professor started out the door, she turned around and finally asked, "Oh, by the way, how do you go to sleep at night?"

The child's father perked up. He seemed to believe she was onto something. "Show the lady, son. Show her how you get to sleep at night," he said.

The child lay down. He extended his right arm, draped it over his head, then inserted his index and middle fingers upside down into his mouth. My classmates roared. Those in the evaluation room heard us and turned to look at the mirror. This angered our professor.

My classmates and I returned to our classroom for further instruction. The professor came in and announced that I was a "witch." I had used some kind of witchcraft in my prediction, she explained. She pursued it no further. And she would not allow me to pursue it either. She would not accept that I had any scientific basis for my observations. She had little to do with me after that.

I took on extra therapy at California State. The professor in charge of clinical experience had me treat the cases of any speech interns who called in sick. I involved the parents and pointed out habits that I saw. After all, therapy was to be whatever we could come up with that worked. There were no pat answers. Professors did not teach a method but required that interns develop their own. I got into trouble, however.

When the other interns resumed their cases, the parents asked why they hadn't pointed out the habits. This made the other interns mad. I was told not to use my own methods anymore.

My first job was at the Torrance, California, school district in 1959. I tried the traditional methods during the first year, and, as I explained in the introduction of this book, I was disturbed by the unpredictability of the results. School officials finally permitted me to do some research. In 1962, I began a two-year controlled study with 162 kindergarten-aged children of very mixed ethnic backgrounds in an effort to determine the validity of current methods.

All children were professionally screened for psychological, emotional, dental, and speech problems to determine possible etiological factors. After two years, the experimental group receiving traditional speech therapy statistically had acquired no more speech proficiency than the control group receiving no therapy.

The study revealed that neither emotional nor dental problems could be linked as etiological factors for speech problems. Emotional and dental problems were still present in the children, however. I reasoned that a third factor, yet unresearched, was the common denominator.

I reexamined the findings and carefully scrutinized every detail of the speech production mechanism, proving conclusively that the common denominator was oral habits. Persistent and consistent oral habits cause most speech problems.

I developed a therapy to correct problems that were caused by oral habits, and within only a few months I corrected the speech of patients whose problems had resisted traditional therapy for two and three years. Other research followed to prove that many dental conditions and other problems also are caused by oral habits.

My findings indicate that since speech is an overlaid function of the oral structure, any oral habits that affect growth and development also affect the speech and language.

As a result of my research, I use dentition and occlusion as diagnostic tools to determine the exact oral habit that interferes with the speech-language production of an individual. That habit is then corrected and the muscle function retrained, thus allowing for successful speech correction. My correction percentage rate is about 98 percent.

In other research I have proven that nearly all TMJ (temporomandibular jaw) and MPD (myofacial pain disorders) syndromes are functional problems caused by oral habits. Today this particular problem affects over half of the population, but a disproportionate number are females between the ages of twenty and forty years. This is caused by a sexual-type habit that is consistent and persistent but can be corrected through the therapy.

Speech Therapy History

Before I started my own research, I delved into the history of speech treatment. I wanted to figure out how the current treatment methods had evolved. It seems that as long as people have spoken, they have wondered at those who were unable to do so.

The first medical description of speechlessness was given in the Edwin Smith surgical papyrus, the most ancient extant of Egyptian surgical text, written in the Old Kingdom (3000–2500 B.C.). In his writings, Homer used two terms to denote speechlessness—anaudos and aphasia-speechlessness—caused by emotion and usually accompanied by lachrymose eyes.

The literature of antiquity contains descriptions of speechlessness caused by physical or emotional injuries. Centuries ago, Aristotle wrote, "Lisping is due to the inability to master a letter. . . .; stammering is due to the dropping out of some letter or syllable; hesitancy is due to the inability to join one syllable to another. All of these are due to the want of power; for the tongue is not an efficient servant of the intelligence." (49:102)[1]

In biblical times, there were instances of defective speech from Moses, who spoke of being "slow of speech and of a slow tongue" (Exod. 4.10) to the Ephraimites whose known inability to pronounce the consonant *sh,* as in *shibboleth,* was used as a test to distinguish them from the Gileadites in a battle at the river Jordan (Judg. 12:6).

Among the many unique features of the Hippocratic corpus is the appearance of the first collection of clinical observations of speechlessness combined with various theories regarding the causes and cures of the symptom.

Hippocrates' books *On the Epidemics* afford numerous examples of cases of speechlessness. The early writers were

1. Numbers refer to the list of sources beginning on page 181.

concerned with prognosis and not the more modern scientific goal of diagnosis. Up through the sixteenth century the field of speech pathology was largely confined to descriptions of cases of speechlessness usually associated with some kind of injury or illness. These early writers did not mention problems of articulation.

Many consider John Conrad Amman the father of speech pathology. His book, *A Dissertation On Speech,* written in the seventeenth century, was important in the areas of phonetics and teaching of speech to the deaf. Amman developed methods, formulated problems and theories, and established standards of analysis that have influenced many important scientists in the history of logopedics. (1:V)

Throughout the eighteenth century and the first half of the nineteenth century, the speech specialists apparently confined their efforts to the care of stuttering.

In America, the influence of foreign teachers and physicians was felt during the entire nineteenth century. Translations of foreign articles appeared with surprising frequency. Methods of teaching found their way across the Atlantic and were adopted by speech specialists who often claimed them as their own.

In 1825, an American, Dr. Christopher C. Yates of New York, invented the "Mrs. Leigh's Method" for the care of stuttering. The rules for this method read as follows: "The stammerer is to press the tip of his tongue as hard as he can against the upper row of teeth, is to draw a deep breath every six minutes, and is to keep perfect silence for three days, during which this pressing of the tongue and the deep inspirations are to be continued without interruption. During the night, small rolls of linen are placed under the tongue in order to give it the required direction even during sleep. . . ." (49:106) Mrs. Leigh's Method is reported to have had a great deal of success even in the worst cases but, unfortunately, it was not lasting.

Around 1841, American surgeons took up their knives to follow the fashion of Europe in the treatment of stuttering. Blood flowed in profusion from useless clipping and slicing of tongues and mutilation of other organs. When it was discovered that the benefits of such surgeries were in no way proportionate with the amount of suffering entailed—since the trouble often returned before the wound had healed—a reaction set in, and the most prominent surgeons admitted that treatment by operation was worthless.

The mania for operations had hardly passed when mechanical "cures" came into vogue. They were many and varied: linen under the tongue, pebbles or cork between the teeth, and many other peculiar contraptions. The most popular appliances, invented by Robert Bates of Philadelphia, were patented, widely advertised, and sold for $35 each.

Early in the nineteenth century, Alexander Bell was recognized as an authority on pure diction and speech defects. His book, *Stammering and Other Impediments of Speech,* published in London in 1836, repudiated the wanton cruelty of surgical operations, the ministering of the quack, and the persons who have not studied the phenomena of speech. (49:11)

In the mid-1840s, Alexander Melville Bell, in his books *The Observations of Speech, The Care of Stammering,* and *The Principles of Elocution,* proclaimed that the cause of stammering originated in infancy—either from bad example or misdirection, or indirection of the early attempts to articulate. Since stammering resulted from a bad habit, Professor Bell concluded it was "therefore beyond the province either of medical or surgical treatment, and, within that, exclusively, of the educator." (49:111)

Melville Bell developed a system of alphabetics, known as visible speech. Professor Bell's main purpose in developing visible speech was to evolve a means of writing any language phonetically, and thus make it possible for anyone familiar with the symbols to pronounce words in any language, whether he had ever heard them or not.

In 1860, Melville Bell and his brother David Charles Bell published *Bell's Standard Elocutionist.* It is claimed that the book did more than any other agent on two continents to "make speech articulate." (49:112)

Alexander Graham Bell, who had worked with his father demonstrating visible speech, opened a school of vocal physiology in Boston in 1872. He announced that the school was for the correction of stammering and other defects of utterance and for practical instruction in visible speech. (49:112)

In 1875, at Boston University, Alexander Graham Bell advertised a "class for parents and teachers of those with defective speech." He gave instruction in the mechanism of speech with special reference to the correction of stammering and other forms of defective speech. This was probably the first university class for the instruction of teachers of speech correction. (49:113)

The Bell methods of instruction and visible speech were adopted in many parts of the country. In 1881, in *The Voice,* a magazine devoted to voice culture, Benjamin Cutter wrote: "Visible [s]peech, by teaching the pupil the sure and exact modes of organic procedure, cannot fail to obviate the difficulty of clear enunciation. Visible [s]peech is being introduced into private schools and has been endorsed by many state and city schools, superintendents, and educators." (49:113)

In an attempt to correct the articulatory defects of pupils, Lewis B. Munroe included in his fourth reader a diagram of the organs of speech and a referral for teachers to Bell's visible speech. Munroe suggested that the remedy for speech defects was "to substitute a conscious, intelligent use of the organs for instinctive action which is natural" (49:114)—thus the use of intelligence to compensate for an articulation disorder.

Before the twentieth century few services existed for individuals with speech, language, and hearing disorders, except for several special schools for deaf children and the occasional efforts of individual physicians and teachers. In Europe, the early growth of the profession was closely allied to medicine. In America the initial impetus came from speech, education, psychology, and medicine. A separate professional identity emerged in the 1920s, when speech and hearing services were initiated in some public school systems; university programs were developed to prepare clinicians and researchers in speech-language pathology and audiology; and a national organization, now the American Speech and Hearing Association (ASHA), was founded.

By examining the growth and development of ASHA, a clear picture emerges of the development of the speech profession in America from the 1920s to the present.

A hint at the state of the profession at the time ASHA was founded can be observed by noting the backgrounds of the twenty-five charter members. Only two of them held doctorates in speech with concentration on speech pathology. One had a Ph.D. in psychology; one held a Ph.D. in speech with emphasis on phonetics. Three founders were M.D.s, two of them specialists in otolaryngology. Several founders had majored in general speech and later developed their knowledge of speech disorders through experience rather than academic training. Three charter members were English teachers.

A further dating of the beginning of the profession may be observed by the fact that the *Quarterly Journal of Speech Education* deemed it useful to print an article titled "How to Begin Speech Correction in the Public Schools" by Walter R. Swift in 1919.

During the first twenty years of ASHA's existence, the primary professional focus of its members was on stuttering. Martin F. Palmer, president of ASHA in 1948, wrote about his earliest recollections of the organization:

> The Association seemed to have been created to permit annual arguments about stuttering. Other disorders received mention from time to time, but the delegates in the halls, and in the official sessions, were busy proving that they: 1) knew the cause; 2) knew the treatment; and 3) everyone else was ignorantly misguided.
>
> There was a full range of opinion from the lefty-righty pat-a-cake of the cerebral domenists to the rolled up eyes and averted gazes of the psychological misfits. Former (and not so former) stutterers could be stepped on anywhere. (38:323)

At the first thirty ASHA conventions, delegates presented approximately 200 papers on stuttering, almost twice as many as in any other category. The *Journal of Speech and Hearing Disorders* published four times as many articles on stuttering than on any other type of speech problem during its first twenty years.

The 1930 convention devoted the entire three-day program to stuttering, with thirty of the recognized authorities in that field appearing in a symposium. Each stated his theory as dogma. One-third of the participants were doctors of medicine; only four were doctors of philosophy. At that time, very few graduate programs in speech pathology existed. Until about 1920, speech disorders, especially stuttering, had been widely assumed to be medical problems, which accounts for the large number of specialists with medical backgrounds.

Since 1940, ASHA has grown in size and stature. It has more than 33,000 members, a permanent national headquarters, and a number of full-time employees. No real clinical progress has been made in diagnosing the cause of or in correcting articulation disorders, however. The correction rate for articulation disorders is the same now as it was when ASHA was founded. No major breakthrough has occurred in the correction of either articulation disorders or stuttering. Although many have had their pet theories about the cause and correction of stuttering, the stutterer can expect no more relief from this problem today than he could have expected in 1920, or in Hippocrates' time for that matter.

Thanks to the standards set by ASHA, today's clinicians are better educated than the "speech teachers" of the 1920s. But when one considers that the material being taught in today's colleges prepares the clinician for a success rate no better than that of the speech therapist of the 1920s, it is difficult to say that the clinician is better qualified. In fact, many colleges have lost the support of charitable organizations for their speech therapy programs because they could not prove the effectiveness of their methods. For some colleges, this meant the forced closing of speech departments; others closed voluntarily. Many kept running with government funding. Now government funding for speech services is being pulled, and many therapy jobs have been lost.

Speech Therapy Today

The lack of success in treating articulation disorders led to the therapists redirecting their focus. They turned to developing language programs for disadvantaged children. This abandonment of speech correction for language programs was made possible by reallocating funds previously earmarked for failing therapies.

Current research testifies that articulation therapy provides no better than a 35 percent success rate. Research also shows that self-corrections are about equal to therapy-driven corrections. Many school administrators working with restricted budgets say, "Let's not provide speech therapy until a child is in fifth grade, because self-corrections will occur in many."

This has allowed for the shift to language therapy for minority students entering school districts. Many minority students have low language skills, so it appears reasonable to use available funds to aid them.

I now know that these children need speech therapy to enable them to learn a new language. With my therapy, therapists can alleviate the difficulties these children have with the sound production of their native tongues. This would allow foreign students to acquire English at a very rapid rate and allow for language integration at a much faster pace. By not knowing how to treat causes, the speech field has not been able to aid the education of minority and immigrant groups.

This will be turned around if my discoveries are accepted and implemented by the speech profession. If another discipline provides my therapy, then speech therapists should start looking for classroom assignments as language teachers. There is an irony in this suggestion, though. I found that language and learning problems are the result of articulation problems, which speech therapists now can begin treating.

Today's speech therapists must understand that any success they've had with compensatory measures may have helped them and their students feel better, but they actually taught a handicap. For the students to control their speech, they are handicapped by about 15 IQ points. This I explained in earlier chapters. Today's therapists and old-time therapists are responsible. We didn't know any better.

Many therapists do not specialize in language but focus on other areas such as voice and hearing problems. These specializations are not excluded from my message. All conditions treated by speech therapists have a cause that we can eliminate.

No child should leave kindergarten without the benefit of this therapy, now that therapists can help children reach their potential. Therapists trained to provide my therapy can guide children into developing better bites, correct jaw movement, good speech, and better reading and language skills. The result will be more emotionally adjusted children who can reach their potential. The result will be therapists healed from "profession depression," a concern expressed by Darrell Dern, president of the Arizona Speech-Language-Hearing Association, in a September 1999 newsletter.

Dern addressed his disbelief over the loss of speech pathology jobs after a change in government funding. He wrote, "What I thought would be a 'forever' profession was turning into a 'no longer' profession. I started questioning my commitment to the profession. . . . Then it happened. As I started to talk to my colleagues about my feelings, I found that *I was not alone!* My goodness, I wasn't the only one feeling depressed and disinterested. I discovered a new illness. I have come to call this illness 'profession depression,' and it is marked by all the symptoms of post traumatic stress syndrome. So what's a professional to do? . . ."

Chapter Eleven

History of Dentistry

I developed an interest in dentistry when I discovered how oral habits affect dentition. Seeing how habit patterns shape the oral structure fascinated me. I marveled how the oral structure restores itself when damaging habits or improper dental treatments are discontinued.

Having made such a discovery, I wanted to understand current dental practices. These I learned from Dr. Arthur Berke, a children's dentist, as we closely observed changes taking place with my therapy. As I mentioned in earlier chapters, we found ourselves opposing some current dental principles and practices.

I became very curious as to how these principles and practices evolved, so I studied the history of dentistry. In 1982, I wrote the following overview as background information for my dental and speech seminar students. I include it here for readers who are interested in the background of dentistry and how it contributes to current practices.

A Historical Overview

Tooth wear and periodontal disease, as well as skeletal structural deformities, have always existed in man. This is evidenced by the embalmed bodies of the Egyptians of 4,000 years ago. At that time periodontal disease was the most common of all diseases.

The Chinese in 2637 B.C. wrote about disease of the mouth, toothaches, and other dental and gingival diseases. It is interesting that they inserted their medicines for the relief of toothache into the ear or the nostril. They also thought that worms produced caries of the teeth.

It is said that a Japanese native dentist who traveled the country was an expert in extracting teeth with only his fingers as instruments.

Archaeological finds have shown that the Etruscans and Hindus in about 2000 B.C. used artificial teeth made of wood or wood covered with gold and held in position by gold or silver wire. Evidence found on the slopes of the Andes tells us that the Incas filled the teeth of their royalty with gold inlays and even performed implants. Archaeologists have described this work by saying that the upper middle teeth were decorated by the insertion of thin gold disks in cavities drilled or bored in the enamel upon the face of the teeth. They also noted that the right central incisor did not belong to the jaw but was transplanted to replace a lost incisor.

The Ebers' papyrus contains many references to gingival disease and prescriptions for strengthening the teeth. It also refers to specialists in the care of the teeth. The practice of dentistry historically can be traced back to Egypt in about 3000 B.C. and a man named Hesi-Re, who was with King Zosh's "chief of the toothers and physicians." (29:78)

Although the Egyptians made wonderful achievements, they never seemed to acquire the ability to generalize.

The Greeks used the Egyptians' knowledge and applied their own scientific inventiveness and causative thinking. This later became a part of western civilization's knowledge of medical science.

The Greek physician Hippocrates discussed in his writings the functions and eruption of the teeth and also the etiology of periodontal disease. He responded to complaints of dental problems by creating a dentifrice to clean teeth and a mouthwash. In his writings, Hippocrates made early reference to TMJ disorders. He ascribed systemic disturbances to irregular teeth and connected them with skeletal peculiarities of the skull as shown in his works on *Epidemics* (sixth book) where he says, "Among those individuals whose heads are long in shape, some have thick necks, strong members and bones, other have strongly arched palates, their teeth disposed irregularly, crowding one on the other, and they are molested by headaches and otorrhea." (8:112)

Both the Greeks and Romans showed an appreciation for not only the usefulness of teeth but also of the teeth as a contributing factor to the beauty of women and correct enunciation in oratory. The first code of Roman law, written in 450 B.C., prescribed special penalties for knocking out the teeth of an individual, either free man or slave, to the extent that if a master or his agent knocked out the front teeth of a slave, that slave was entitled to immediate freedom.

The first extensive Roman work on medicine was written by Arueluis Corneluis Celcus, the historian of scientific medicine of his time, in about A.D. 24. In Book VII, Chapter XII, entitled "Operations Requisite in the Mouth," he says, "And if one tooth should happen to grow in children before the other has fallen out, that which should have dropped must be scraped round and pulled out; that which is growing in place of the former must be pushed into its proper place with the finger every day, till it comes to its own size." (8:112)

Galen advocated this same procedure in his writings of about A.D. 180. He also said that a tooth that projects beyond its neighbors should be filed off to reduce the irregularity.

Because of various conditions, largely connected with the Church, Galen monopolized medicine for more than a thousand years, and there was little advance in European medicine. Among the Arabians, however, medicine continued to advance, and we find references to principles of modern orthodontia.

With the dawning of the Middle Ages and the end of the Dark Ages, advancement began again, stimulated largely by the writings of the Moslem physicians of Arabia, Persia, Egypt, and Spain.

In the early sixteenth century, Fallopius and Eustachious conducted the first study on the anatomy of the jaws and teeth. They described the innervation of the skull, the hard and soft palates, and the anatomy and development of the teeth at every state, from early uterine life until eruption of the permanent teeth.

The techniques of dental practice and the study of dental anatomy remained distinct until 1562 when Ambrose Pare, a Parisian barber surgeon, began to study methods of operating and repairing cleft palates. Up to this time, no one group or individual dominated the realm of dental knowledge. European barbers, blacksmiths, and barber-surgeons performed extractions and fillings and created simple bridges.

The dental anatomy was still considered part of the physician's domain. One must remember that, at best, until 1700, dental treatments were confined to the aristocracy and were intended to improve the appearance of the wealthy.

Then in 1755, Pierre Fauchard, a French surgeon and scientist, took a major step toward expanding dental care to the general public. He separated and compiled dental knowledge into a comprehensive book covering every then known phase of dental practice. Although Fauchard's patients still included many of the crown heads of Europe, this expansion of knowledge eventually led to better dental care for the masses.

Up to this point in time, throughout history, most civilizations practiced some form of dentistry. But each civilization had its own methods for doing so. There was little or no exchange of ideas, and, with the possible exception of the Greeks, one civilization did not build on the knowledge and experiences of another.

It was not until the eighteenth century and Pierre Fauchard that the continuous history of dentistry as we know it actually began. Hence, he is called the "founder of modern dentistry."

In order for improvements to be disseminated throughout a profession, however, there must be free access to and discussion of any new ideas and practices. Unfortunately secrecy shrouded eighteenth-century dentistry in Britain and France. Would-be dentists purchased the right to study under a preceptor, but they did not discuss the "trade secrets" for fear someone else would learn for free. Furthermore, there was more money to be made by being the best dentist in town, so they were slow to show any new discoveries to competitors. Only when "competitors came to be regarded as 'colleagues' was there an emergence of a true 'profession.'"

In the free atmosphere of the young United States, European secrecy was more difficult to maintain, and, though it took 200 years, it was finally abandoned. (9:18)

In the early American colonies, dentistry was practiced by goldsmiths, jewelers, ivory turners, umbrella makers, blacksmiths, mechanics, wigmakers, tinkers, engravers, barbers, and itinerant jacks-of-all trades. All but a few physicians refrained from rendering dental services. There were no dental schools, no diplomas, and no state regulations. Anybody who so desired could set himself up in practice with no more than the opening of a blacksmith shop or a barber shop. It has been said that in the colonial period, a tooth was pulled or broken off for the relief of a toothache, and strength was the only operative requirement. Anyone who could make false teeth was free to do so and to fit them in his own way into the mouth of anyone to whom they could be sold.

A look at the history of dentistry in colonial America gives one the feeling of a lack of continuity. The European dental practitioners were largely uninterested in setting up shop in the colonies. Therefore the early settlers on this undeveloped and sparsely inhabited continent were forced to solve their own dental difficulties. The result was that each dentist seemed to be doing his own thing in his own way without much communication between the practitioners.

The first European dental practitioner, William Dinely, arrived in the Massachusetts Colony in about 1636. Then in

1734, James Mills announced in the *New York Weekly Journal* that he was continuing the practice of James Reading "at his shop in the House of the Deceased." In 1736, he advertised again, stating that he safely and carefully extracted teeth, having been instructed in the art by "the late James Reading, deceased, so famed for drawing teeth." (29:79)

It is evident, therefore, that between 1636 and 1734, the number of European barber surgeons or "tooth drawers" in America had increased. They were few in number, however. They practiced mainly in the towns, not venturing into the rural and more isolated areas. Their air of secrecy prevented most of them from sharing their information with American practitioners.

In 1755 and 1761, the first American surgeon-dentist appeared in Virginia and in Boston. In 1768, Paul Revere of Boston advertised his services as "business of a dentist." From 1775–1800, forty-four others advertised and offered dental services with the title "dentist" attached to their name, or they simply called themselves dentists. (46:31)

Many of these early Americans made some important contributions to the field of dentistry. Paul Revere, whose dental patients included General George Washington and General Joseph Warren, first introduced forensic dentistry in America. General Joseph Warren, a physician and leader of the colonists, was one of the first casualties at the battle of Bunker Hill. The dead were hastily buried, but a year later, Massachusetts wished to honor General Warren and exhumed the remains. They had great difficulty in identifying the body, however. Paul Revere identified the body of his patient because of an artificial dental appliance that he had constructed for Warren.

In 1769, John Greenwood, a highly skilled dentist, made the first of a number of dentures for President George Washington. Greenwood was the first to employ spring clasps for the retention of partial dentures in 1792. He was proficient in all types of dentistry and is reported to have been the first to use a foot drill, which he made from an old spinning wheel belonging to his grandmother.

Even more significant, he was an early advocate of regular dental checkups for children. Addressing parents in the *Daily Advertiser,* New York, July 20, 1790, Greenwood wrote of "the attention that is necessary to be paid to children's teeth at the time of shedding and after, being of so great importance to their regularity, evenness, and future preservation." (14:9)

Josiah Flagg Jr. was born in Boston in 1763. He became interested in dentistry because of Paul Revere, who had taught him about metals being used in dentistry. Flagg pioneered root-canal therapy. In the Federal Orrery, Boston, November 30, 1795, Flagg announced that the sensation or nerves of the teeth in the head can be extracted by a "simple, safe, and easy process with instruments, and the teeth still preserved and prevented from further decay." (14:9) Up until that time, the standard procedure was to destroy the nerve with knife, acid, or cautery.

In commemoration of Flagg, the South Carolina State Dental Society erected a monument that reads: "In this churchyard is buried Josiah Flagg, 1763–1816, first native-born American to make dentistry his life's work, and the first to carry to any foreign land evidence of American dental progress. Soldier in the American Revolution, soldier in the War of 1812. Died in Charleston, September 30, 1816." (29:82)

During this period Europeans had some influence on American dentistry. Michael Poree, "Operator of the Teeth" from Paris, published the earliest known American article on dentistry for the information of laymen. In the *New York Gazette* and *Weekly Mercury*, December 11, 1769, Poree warned against "erroneous operators for the teeth" and reminded his readers that "it is absolutely necessary for us to have the number of teeth which was given to us . . ." (29:80)

John Templeton, a surgeon dentist from Europe, addressed himself to the dental care of children. The Boston Medical Society certified Templeton for proficiency in his profession. He appears to have been America's first orthodontist, undertaking the "Charge of Children's Teeth" as soon as they begin to "shed" and the correction of "Irregular Teeth" in a practice that occupied the 1780s. (29:80)

As early as 1792, the medical profession in this country had acknowledged the value of a dental service in a healthcare institution when the New York Dispensary, founded in 1791, conferred upon Richard Cortland Skinner, surgeon-dentist from London, an official position in its organization. This resulted in the first dental clinic in America. R. C. Skinner also wrote the first dental book published in this country, *A Treatise on the Human Teeth*. It was written about 1794 and copyrighted in 1801. Historians agree he is the founder of American dental literature.

By the beginning of the nineteenth century, dentistry was in the hands of those calling themselves dentists. Dentists were assuming an identity as possessors of special knowledge. The host of physicians, barbers, blacksmiths, schoolmasters, or "tooth drawers" who performed dental services were differentiated from those dentists who claimed dentistry as a sole occupation instead of a sideline. The dentist also claimed the greatest expertise in dental matters.

A phenomenon related to the transfer of knowledge and establishment of a professional dentist-identity occurred around 1800 with delineation of dentistry as a profession by attacking the barber-surgeons and tooth drawers as quacks and charlatans.

In the nineteenth century, the field of dentistry seemed to verily explode. The advancements of the field as a profession as well as the discoveries within the field were so numerous that only a few of the highlights are mentioned here in chronological order, not in order of importance.

The Articulator

J. B. Gariot of France felt that to ensure the function of artificial teeth, they must be set in proper alignment. In 1805, he designed the articulator, a device for holding in what he thought was proper alignment, the plaster casts to which artificial teeth are mounted before installation in the mouth. "The articulator reproduced the chewing motions, although not completely." (12)

First School to Teach Dentistry

The Baltimore College of Dental Surgery, founded in 1840, is known as the world's first dental school. But few are aware that an earlier school was an indirect ancestor of the "world's first."

From 1825 to 1830, John Harris, M.D., D.D.S., conducted a school of medical instruction while engaged in the practice of medicine and dentistry. His educational requirements, his preparation for instruction, lectures, and establishment of a dental clinic marked it as the first American school of instruction on dental surgery.

Of the students who took Dr. Harris' course, all but one took up the study of dentistry. His students included his brother, Chapin A. Harris, a cofounder of the Baltimore College of Dental Surgery.

Another of his students, James Taylor, established the second dental college, the Ohio College of Dental Surgery in Cincinnati.

These alumni of Dr. Harris' school also edited the first two dental publications in this country, the *American Journal of Dental Science* and the *Dental Register of the West.*

Dr. Harris' professional attainments, his early realization of dentistry as a health measure, and his pioneer efforts in dental instruction have distinguished him as the "father of dental education."

Institutionalization of Dentistry

The process of institutionalization in dentistry could be considered to have had its start in 1839. It was then that the first dental journal, the *American Journal of Dental Science,* was established. The first national dental society, the American Society of Dental Surgeons, was organized on August 18, 1840. The objectives of this society were to promote unity and harmony among dental surgeons, to advance the science by free communication and interchange of sentiments, and to give character and respectability to the profession.

The First Dental School

The first dental school, the Baltimore College of Dental Surgery, was founded on February 1, 1840. This was an important step in the professionalization of dentistry. For the first time, men could receive thorough training of high quality, explicitly for the practice of dentistry, whereas previously the best qualified dentists had been through only two-year apprenticeships. In 1845, the second dental school, the Ohio Dental College, was established in Cincinnati. These two schools widely publicized their excellent training opportunities and slowly the idea of formal dental education came to dominate the profession. (62:32)

First Dental Law

The first law in the United States regulating the practice of dentistry was enacted in 1841. The legislature of the State of Alabama made provisions within the state's medical boards for practicing dentists to "examine and license persons to practice." (25:17)

Dental Hygienists

Dr. Alfred C. Fones of Bridgeport, Connecticut, a follower of Dr. D. D. Smith of Philadelphia (who was an early advocate of the techniques of prophylaxes), established the first formal school of dental hygiene. On November 17, 1913, thirty-three women, including school teachers, trained nurses, experienced dental assistants, and the wives of three dentists, began the first course for dental hygienists. The first licensee law was passed by Connecticut in 1917. The American Dental Hygienists' Association was formally organized on September 12, 1923.

Orthodontics

As stated earlier, Hippocrates (460–377 B.C.) said, "Among those individuals whose heads are long in shape, some have thick necks, strong members and bones, others have strongly arched palates, their teeth disposed irregularly, crowding one on the other, and they are molested by headaches and otorrhea." (8:112) This was possibly history's first recorded orthodontic diagnosis.

Around 40 B.C., the Etruscans used wires to straighten teeth. Aureluis Corneluis Celsus (258 B.C.–A.D. 50) recommended pushing teeth into place with the fingers. Around the same time (A.D. 23 to 79), Plinius was perhaps the first to treat dental irregularities through the use of instruments. He condemned extraction and advocated filing elongated teeth to bring them into proper alignment. He stated, "Teeth are not necessary for mastication alone, for the foremost ones regulate the voice and words, producing by the beat of the tongue special sounds." (17:3)

In 1805, Gariot, in his "*Treatise on the Diseases of the Mouth,*" cautioned parents to give attention to the first dentition in order to secure regularity in the second dentition. But, he said, if irregularities did occur, they could be corrected either by radical operations or by expectant treatments. The radical operation was rapid and painful and was done by spreading the neighboring teeth with ligatures to make room for the new positions of the irregular teeth. This operation required a great deal of skill to avoid permanent injury and was not recommended. In the expectant treatment, the appliances used were ligatures, either attached to neighboring teeth or to a gold on ivory plate.

Historically, until the nineteenth century the most widely used method for straightening teeth was extracting the teeth and replacing them with dentures. Then in the nineteenth century some believed in the importance of preserving the teeth by filling and straightening them with appliances.

The controversy raged between those dentists called regulators, who favored the straightening of crooked teeth, and the dentists called extractionists, who favored extracting teeth and replacing them with dentures. Benjamin James of Boston favored preserving the teeth. In 1814, in his "*The Management of the Teeth,*" he stated, "The teeth are so important in dividing our food, as well as conversation, and are besides so highly ornamental, as to render it a positive duty with everyone to study the causes which lead to their premature destruction."

In 1817, Serres, in his "*Essay on the Anatomy and Physiology of the Teeth,*" described the tooth width-arch length discrepancy theory when he discussed the causes of irregularity in the second dentition. He believed that this was chiefly due to the extension of the jaw being insufficient to accommodate the larger and greater number of teeth in the permanent dentition. This view was consistent with the extractionists' theory.

In 1840, DeLoude first mentioned the theory of inherited tendencies as an etiologic factor "where the child inherits the jaw of one parent and the teeth of the other."

During this time, many offered their theories on the irregularities of the teeth. Lefoulon, in 1839, was the first to name the art of regulating teeth "orthopedia dentaire" and later, in 1841, "orthodontosie." On the subject of regulating teeth, Lefoulon stated, "To extract is not to cure but to destroy." In interpreting the management of second dentition, he discussed the pros and cons of the extraction of the deciduous teeth and concluded that the "every day's experience shows that the teeth, which at first to a trifling extent are irregular, will straighten themselves as the circle of the jaw expands. We would urge upon our brethren to take a lesson from Nature, who thus established harmony for a moment disturbed." (60:474–475)

As to the causes of irregularities of the teeth, his ideas are based upon biological phenomena controlling growth, form, and dimensions of organs and tissues, including prenatal conditions, producing disturbances in the size of the deciduous teeth.

In 1846, James Robinson of London, in his *The Surgical, Mechanical, and Medical Treatment of the Teeth,* devoted four chapters to a consideration of irregularities of the teeth. He classified irregularities of the teeth as temporary and permanent. Temporary, he said, arose from too rapid advancement of the teeth before a corresponding resorption of the first had taken place. He believed that permanent irregularities were caused by the increased size of the second teeth as compared with the first and by their eruption before sufficient expansion had taken place in the jaw.

In 1864, Ballard developed the theory of "fruitless sucking" as the main cause of irregularities. His interesting and keen observation had little influence upon those interested in the development of orthodontia, because it is not found recorded in subsequent literature.

Isaac B. Davenport, in 1887, strenuously opposed extraction "for besides the loss of important organs we shorten the bite, contract the mouth, disturb the facial expression, and secure at least an arrangement of the teeth less favorable for their preservation than existed before." (61:211)

The most important advancement in the history of orthodontia came in the works of Edward H. Angle, M.D., D.D.S. In 1907, in his seventh edition of *Malocclusion of the Teeth,* he stated: "Occlusion is the basis of the science of orthodontia." His conception of occlusion completely changed the study and treatment of malocclusion of the teeth. He classified occlusion based upon the mesiodistal relationship of the dental arches. Dr. Angle's unflinching and uncompromising position against the extraction of the teeth in orthodontic procedures rendered an inestimable service.

Calvin S. Cass, M.D., D.D.S., advocated the extraction of teeth for the correction of malocclusion, and until his death led the group of men who believed in this practice. He felt it necessary to extract premolars for the correction of biprotrusion. He showed interest in radicular movement and pointed out methods to produce it. He was among the first to apply intermaxillary elastics.

The annual meeting of the American Society of Orthodontics in 1911 was the scene of bitter and diverse open debate. One of the two strongly polarized groups was led by Drs. Cass and Cryer. They were known as the "rationalists" or, in today's language, the "nonexpansionists," and they held that the extraction of premolars in orthodontics was an accepted approach.

The other group, led by Drs. Angle, Dewey, and Farris, was known as the "new school," or now, the "expansionists." This group held that there was no reason for these extractions.

According to the rationalists, a majority of cases of malocclusions requiring extraction were caused by genetic traits. The new school contended that factors of heredity have nothing to do with occlusion, and that nature's process is to develop a perfect occlusion in everyone if left alone to do so. The first group quoted Darwin, Mendel, and Huxley. The second quoted the Bible.

The rationalists held that the alveolar process follows the teeth when moved, but interstitial bone would not grow beyond its inherited limits. Occlusion should dictate facial harmony.

The expansionists believed that interstitial bone could be stimulated to grow to any dimension and to any form and direction desired by the proper application of force. Thus extraction, to them, was malpractice. Occlusion had little to do with facial harmony.

Dr. Mathias observed that this argument still brews today in myofunctional therapy and other disciplines of dentistry. He believes that the contentions of both schools are still theories, and, as such, neither can be proved in the laboratory, even today. It is also worth noting that our present debate in educational circles in Arkansas, Louisiana, California, and other states has proved the controversy of 1911 is not over yet.

Serial Extractions

Today there is a broad-based diversity of opinion concerning extractions. This could be due to a lack of complete understanding of orofacial growth patterns, as well as the effect of orthodontic intervention on such growth. (4:22)

Back in 1858, Kingsley stated: "It is surprising the vast amount of culpable ignorance daily and practically displayed by the profession, even in regards to the development of the permanent teeth. . . . Removing temporary teeth to make room has been the cause of more irregularity than Dame Nature was ever guilty of." (4:481)

Seventy years later, in 1929, Kjellgren, the originator of the term "serial extraction," described the spontaneous improvement of alignment of the mandibular permanent incisors when primary canines were spontaneously prematurely exfoliated. Since then the use of serial extraction procedures has been developed and described by many authors. (4:21)

According to Robert E. Binder, D.M.D., "Serial extraction procedures as currently applied by dentists are not truly preventive in nature. They are rather a series of related and correlated steps taken in an attempt to interrupt a developing malocclusion of the mixed dentition before it can be fully expressed. The intent of serial extraction procedures is the removal of primary and secondary dental units in a sequence that will lead to proper (ideal) inter- and intra-arch occlusal and functional relationships of the remaining teeth without the use of appliance therapy. At the other extreme, serial extraction procedures may precipitate a more complex orthodontic problem than existed originally, then requiring compound and complex orthodontic appliances.

"For the most part, serial extraction produces a result which rests somewhere between these extremes. When any type of interceptive or preventive treatment is undertaken in the mixed dentition, it must be stressed that further orthodontic intervention may be required to produce a more than minimally acceptable result." (4:21)

TMJ

A review of the current controversy over the treatment of temporomandibular joint (TMJ) syndrome and myofacial pain dysfunction (MPD) syndrome reads something like the story of the practice of dentistry in the American colonies. Everyone is doing "his own thing" in "his own way" without much communication between the practitioners. Today's TMJ practitioners include dentists, medical doctors, psychologists, psychiatrists, physical therapists, and myofunctional therapists. Each has his own theories as to the etiology of the syndrome, and each has his own treatment program for its relief. While surely each one is highly skilled in his own field, there is much disagreement as to which field of practice the TMJ and MPD syndromes belong. Because none of these practitioners has arrived at the true etiology of TMJ or MPD, they are still chasing symptoms with a wide variety of contradicting theories and treatment programs.

In 1918, Prentiss recognized that the loss of molars and premolars produced a distal condylar movement that resulted in direct pressure on the eustachian tube and the ear structures, or impingement of the auriculotemporal nerve. He concluded that a vertical collapse of the occlusion was responsible for the TMJ dysfunction syndrome. (32:438)

In 1920, Wright agreed, stating that the retrusion of the condyle could cause resorption of the tympanic plate, constriction of the canals, and irritation of tympanic structures including the tympanic artery and the chordatympani nerve, causing partial or total hearing loss. (32:438–444)

In 1934, James Costen, M.D., reported that the loss of posterior teeth resulted in overclosure of the mandible and consequently resulted in pressure of the condyle on the retrocondylar structures. Pressure on these structures, mainly the auriculotemporal canal and nerve and the external auditory meatus, resulted in a range of symptoms including impaired hearing, tinnitus, dull pain in and around the ears, dizziness, and alleged sinus symptoms.

Some authors supported Costen's theories, but others disagreed.

In 1951, Zimmerman wrote, "It is evident that from an anatomical and functional point of view there is only one group of symptoms in Coston's syndrome of mandibular overclosure that has a basis of acceptable fact, trigeminal and occipital neuralgias. All other symptoms are questionable." (32:438–444)

Sicher questioned tympanic irritation with mandibular overclosure. He thought that the pain arose from a spasm of the muscles secondary to malocclusion. He believed that the dental malocclusion was the key factor and that the edentulous patient would not have this problem.

In 1937, Schultz introduced the concept of hypermobility caused by "lax ligaments." He was concerned with clinically observable joint disorders such as clicking and dislocation. He believed that an excessive range of condylar movement caused the symptoms of pain and clicking. He directed his attention to the joint itself and emphasized treating the joint rather than the occlusion.

Travell and Rinzler called painful areas within muscles "trigger areas." They pointed out the existence of syndromes associated with trigger areas within the muscles coupled with pain, spasm, tenderness, and dysfunction.

Schwartz felt that psychological and physiological predispositions were more important than the precipitating factors themselves. He thought that patients were predisposed by constitution and temperament to have pain-dysfunction syndrome as a result of muscle overextension or overcontraction. Schwartz considered abnormalities of the occlusion to be the only contributing factors.

In 1969, Laskin offered a revised view of Swartz's TMJ pain-dysfunction syndrome. He felt that, while mechanical factors related to occlusion may sometimes cause TMJ syndrome by muscular overextension or overcontraction, muscle fatigue was the most frequently encountered cause of such spasm. He said that the fatigue appears predominantly related to psychologically motivated, persistent, tension-relieving oral habits, such as clenching or grinding of the teeth. He felt it was functional in nature but with psychological implications.

Some authors believe that inadequate dentition and unsatisfactory occlusion are the most frequent causes of MPD. Others believe that TMJ disturbances are usually related to dysfunction of masticatory muscles and/or emotional disorder.

Histopathologic investigations have attempted to correlate myofacial pain and dysfunction with the direct overloading of muscles or with indirect disturbances of the blood supply, in which cells are deprived of glycogen and oxygen and receive an increasing lactic acid concentration.

Thomson states that the type of occlusion and the number of missing teeth do not contribute to MPD. He believed that MPD has a neuromuscular origin.

An increasing number of clinicians, dentists, and other specialists treat headaches and TMJ disorders by repositioning the mandible with mouthpieces, templates, and bite plates. (These same appliances have been used in treatment for a loss of equilibrium, scoliosis, and lordosis of the spine.) The benefits of such treatment have yet to be proven, however.

Dawson stated that MPD pain can be resolved in most cases in a matter of minutes once the occlusion has been refined, with the exception of traumatized joint surfaces, tendinitis, or damaged muscle. Celenza believed that the TMJ is the last structure to show pathology. "The joints are designed so that they are the top structure of the system. Restoration of harmony between positional guidance components within the adaptive range of that individual is the objective of oral rehabilitation." (32:442)

One school of thought emphasizes the occlusal disharmony as the major etiologic factor in the development of MPD. This theory cannot explain why masticatory pain and dysfunction are comparatively uncommon when occlusal disharmony is so prevalent in the general population, however.

Another school believes that MPD is triggered by emotional tension and may be caused by a combination of psychological stress and muscle incoordination. (32:442)

Weinberg argues that the diametrically opposing etiologic concepts, namely occlusion and emotional stress, are both misused when a patient must be rigidly typecast into one group depending on the views of the clinician. He avoids using the terms "psychosomatic or emotional tension," as these terms imply that TMJ dysfunction is an emotional disorder that can only be cured by psychiatry. (32:442)

Goss believes that the psychological aspect of MPD is an important one, and it has been asserted that without emotional tension MPD does not develop. Lupton supported this assertion by showing that there was no statistically significant difference between a randomly selected group of MPD patients treated with psychotherapy alone as compared to a group treated with only dental therapy. In a selected group of patients with a strong nonorganic component of MPD, it is possible that dental treatment without psychotherapy could do definite harm, according to Upton. Posselt suggested treatment of patients with several MPD symptoms by selective grinding.

Physical therapists believe that there is a relationship between the temporomandibular joint and the spine. Walter Golden, R.P.T., claims that the predisposed person who undergoes postural changes will most certainly have TMJ problems. The diagnosis includes evaluation of leg length variations, curves in the spine, and muscle strength. Various noises in the jaw are analyzed. Golden claims a simple click is relatively insignificant but a double click can represent a dislocated disc. His recommendations for treatment are prescription exercises, bite splints, and manipulative procedures on the jaw.

It is obvious that some large gaps exist in the total understanding of the etiology of MPD and TMJ syndromes, and there is little agreement among the various schools of thought.

W. Wilson, L.D.S., of London, favors the extraction of second molars as opposed to bicuspids and third molar extractions. He pointed out the effect current-day orthodontics has on TMJ when he said:

> A normal, healthy TMJ should be one of the primary objectives of orthodontic treatment. This objective has been unfortunately and sadly overlooked in the past. With the current medical-legal complications, and the number of TMJ patients bringing malpractice cases to the courts today, every practitioner must be able to recognize and treat TMJ pain and dysfunction.

It is of extreme importance in orthodontics that nothing is done in treatment to move the mandible or condyle distally. Of equal importance, nothing should be done to the patient to cause loss of vertical dimension. This could move the condyle into a displaced position off the articular disc onto the weaker fibrous tissue. This fibrous tissue contains the nerves and blood vessels supplying the nutrition of the TMJ.

Pressure on the fibrous tissue will not only cause pain but also will lead to disturbances of the circulation. This in turn leads to degeneration of the disc and fibrous covering of the articulating bones. Pressure can come only from the condylar head.

Because of the above reasons, the following three causes of a pathological position of the condyle in the TMJ must be avoided in orthodontic treatment:

1. Retracted Upper Anterior Teeth

After upper bicuspid extraction. (Retraction can move the condyle into a pathological position and cause loss of vertical by the bicuspid extractions.)

2. **Deep Overbite (Loss of Vertical)**

   A severe loss of vertical dimension occurs with four bicuspid extractions and contraction of the dental arches. The loss of vertical should be avoided for a normal, healthy TMJ.
3. **Class II/Division II-Malocclusion Improperly Treated**

   As these cases almost always have deep overbites, it is essential that bicuspids are not removed causing further loss of vertical dimension. (63:1–3)

In January 1983, the *Journal of the American Dental Association* (JADA) reported on a conference held by the American Dental Association (ADA) on the etiology, diagnosis, and management of temporomandibular joint disorders. The conference concluded that there was insufficient data to permit a comparison of different forms of therapy to establish a priority for their use. The basic principle of using conservative reversible forms of therapy was advocated, however. The following recommendations were made regarding the various forms of treatment used for patients with TMJ disorders:

Occlusal appliances designed to intentionally and permanently reposition the mandible or the dentition should be considered an irreversible form of therapy.

Intentional permanent placement of the mandible in a new position by means of prosthodontic treatment, orthodontic treatment, surgery, or by functional occlusal appliances should be considered an irreversible form of treatment.

For functional disorder, such as MPD syndrome, surgery is not indicated. At the present time, there is no evidence to support the use of surgery for the initial treatment of disc displacement.

For some dysfunctional conditions involving the muscles of mastication, the conference held that the behavior therapy approach is recommended and has reasonable scientific support. The behavior therapy referred to included biofeedback and relaxation training.

In April 1983, the *JADA* printed a letter from Morton Olin, D.M.D., of Arlington, Massachusetts. The letter compliments those dentists who are genuinely concerned for the welfare of their patients. His letter reads:

> The "Report of the President's Conference on the Examination, Diagnosis, and Management of Temporomandibular Disorder" with introduction by Dr. Robert Griffiths helps to restore my faith in dentistry. I read so much so-called "information," particularly on the temporomandibular joint, which is no more than opinion, ill-founded theory, or anecdotal case histories unsupported by appropriate studies, that it gratifies me to see a carefully presented and thoughtfully done report. You may receive a few letters from some members of our dental community objecting to this report for not supporting their own pet theories. To have pet theories is tolerable, but to act on them is potential menace to the public.
>
> I believe that it is imperative that our profession neither rejects nor accepts ideas when they are unproved, but that we scrutinize them critically. Without this kind of scrutiny, we run the risk of appearing more as charlatans than professionals. (22:75–77)

Chapter Twelve

Growth and Development of the Oral Structure

When I was reviewing the history of speech pathology and dentistry, which I presented in the last two chapters, I specially looked for and noted any consideration that was given to the growth and development of the oral structure.

For diagnostic purposes, consider that normal growth and development is that which attains symmetry or balance.

In preparing a handbook for dental and speech professionals attending my seminars, I included some of these observations that other researchers have made. For instance, I pulled material used by H. Harlin Bloomer, Ph.D., in the "Handbook of Speech Pathology," a standard of reference in the field for decades:

1. Osseous defects occur when one or more of the bones of the face develop in a perverted, delayed, advanced, or other asynchronoia pattern of growth. Although each bone has a growth potential that is genetically determined, this potential may be altered by environmental forces such as mechanical interferences and habits of use;

2. Muscular dysfunctions that persist throughout critical periods of orofacial growth can distort the direction of

facial development so that the width of the maxillary dental arch, the inclination of the teeth, or the habitual positioning of the mandible may result in permanent deformities.

. . . In consideration of the influence of muscular function upon bone growth, it follows logically that whereas speech may be affected by the growth of the face and jaws, the habits of speech also can play their part in inflecting growth, an observation that has been made by Froeschels (1933), Green (1937), Palmer (1948), and others. Henry (1937) suggested speech drills as a means of improving occlusion."

Dentists, orthodontists, and myofunctional therapists have for many years believed that malocclusions and tongue thrusts were responsible for many speech problems, and that the correction of malocclusion and tongue thrust would lead to speech corrections. Myofunctional therapists claim a high correction rate when their therapy is used to treat certain articulation disorders due to the fact that they address the tongue's rest posture and the swallowing pattern.

Speech Pathologist and Oral Facial Myologist Roberta Pierce promoted myofunctional therapy in an October 1999 issue of Advance. She pointed out that her tongue thrust patients with speech disorders receive "instruction and exercises to improve their speech, superimposed on a tongue thrust therapy program."

Pierce also gave statistics revealing the lack of attention and training provided speech professionals regarding oral myofunctional disorders. Citing another research study, Pierce reported that 34 percent of surveyed speech pathologists reported that they received no education in this area,

63 percent had no practicum experience, and 87 percent viewed their instruction as inadequate.

Pierce's article was bent on promoting myofunctional therapy to speech pathologists in light of recent funding changes that have resulted in the elimination or downsizing of many speech therapy jobs. She suggests myofunctional therapy as an alternative to be pursued.

But myofunctional therapists overlook two things: First their method does not address causes. They are correct in seeing a relationship between speech and oral function, but what I found is that the relationship is not a cause-and-effect link. Instead speech, swallowing, and tongue disorders have a common cause.

That cause, an oral habit, must be identified and eliminated. But this is not enough. Correct function must be reconditioned. It must be internalized. This brings us to the second factor that myofunctional therapists or myologists have not addressed.

Their therapy does not cure; it does not recondition the reflex system. It is another way of teaching individuals to control or override their habit-conditioned speech and swallowing patterns. Their muscle exercises can only work as long as their patients continue to exert the control. As I explained earlier, this robs individuals of cortical skills needed for language comprehension and concentration.

Over the years, several leading myofunctional therapists have requested information from me. They have asked me to write chapters for their books or to present at their conventions. Unfortunately, I was not ready to present my work. I hope this open-minded group will still consider my discoveries and the training they will need to offer my therapy.

Speech pathologists, who are not trained in myofunctional therapy, also have been trying for many years to show a cause-and-effect relationship between malocclusions and articulation disorders. However, in 1954, Van Riper said that the presence of some anatomical abnormality is of no importance in itself, unless it stands in functional relation to defective speech sounds.

It is obvious that the experts in the fields of dentistry and speech pathology do not agree on the age-old dilemma: Does function follow form, or does form follow function?

I profess that form follows function. Consider that growth starts at conception, and that the growth and development of an individual can be altered or influenced at any time during his lifetime - while in the womb or anytime after birth. Thus, a deviation in a growth pattern can be initiated after conception by a change in function.

Growth and development should achieve anatomical symmetry or balance. For diagnostic purposes, any deviation from a balanced growth is an indication of an interference in the growth pattern.

What can cause a deviation? What could interfere with the growth pattern? If you attended my seminar or have read this book from its beginning, I'm sure you know.

It could be an oral habit. Or metaphorically speaking, it could be the pebble in the shoe.

Sources

Much of the history information in Part Four came from trade journals. Because some journals were used repeatedly, I chose to reference the sources in the text with numbers inside parentheses. The first number corresponds to the following list; the second provides page numbers.

1. Amman, John C., M.D. *A Dissertation on Speech*. Amsterdam: North Holland Publishing Co., 1965.
2. ASHA Joint Committee on Dentistry and Speech Pathology—Audiology, "Position Statement of Tongue Thrust." *ASHA Journal*. Vol. 17, no. 5, May 1975.
3. Barrett, Richard H. and Marvin L. Hanson. *Oral Myofunctional Disorders*. St. Louis: C. V. Mosby Co., 1974.
4. Binder, Robert E., D.M.D. "Serial Extraction in Preventive Dentistry." *Clinical Preventive Dentistry*. Vol. 1, no. 4, July–Aug. 1979.
5. Boucher, Carl O., D.D.S. *Swenson's Complete Dentures*. 5th ed. St. Louis: The C. V. Mosby Co., 1964.
6. Bradel, Sigmund F., D.O.S. "One Hundred Years of Development in Metallurgy and Its Relation to Orthodontia." *Journal American Dental Association*. June 1934.
7. The British Journal of Communication, "Speech Therapy Comes of Age." *The Journal of the College of Speech Therapists London*. Vol. I, no. 1, Apr. 1966.
8. Carranza, Fermin A., Jr., D.D.S. "Prevention, Diagnosis, and Treatment of Periodontal Disease in the Practice of General Denistry." *Glickman's Clinical Periodontology*. Philadelphia, London, Toronto: W. B. Saunders Co., 1979.

9. Casto, Frank M., M.D., D.D.S. "A Historical Sketch of Orthodontia." *Dental Cosmos*. Vol. 76, 1934.
10. Charles, Allan O. "The Leadership of American Dentistry, 1790–1840." *South Carolina Dental Journal*. Fall 1977.
11. "Cradle of Dental Education." *Dental Student*. Vol. 41, Jan. 1971.
12. Dallas, Robert F. "Teeth: Invisible Braces are Planned." *Los Angeles Times*. 1983.
13. Davis, Audrey B. "The Dentist and His Tools." *In the Smithsonian*. 1978. Figures 15–25.
14. Del La Rosa, Federico R., D.D.S. "Evolution of the Conception of Diagnosis in Orthodontics." *American Journal of Orthodontics*. Vol. 37, Jan. 1951.
15. "Dentistry's Hall of Fame." *TIC*. Vol. 35, February 1976.
16. Edwards, Ralph W., D.D.S. "A History of the International School of Orthodontia." *Journal Missouri Dental Association*. June–July 1961.
17. Farkas, Joseph A., D.D.S. "Screen Examination for Occlusal-Mandibular Dysfunction." *Journal of General Dentistry*. Sept.–Oct. 1981.
18. Gilbert, Samuel, G., D.D.S. "Evolution of Present Day Concepts in Orthodontics." *Pennsylvania Dental Journal*. Vol. 27, Apr. 1960.
19. Goldberg, Philip. "TMJ Syndrome: Is It Making Your Head Ache?" *Woman's Day*. Nov. 10, 1980.
20. Golden, Walter W., R.P.T. "Physical Therapy: General Implications for the Treatment of Temporomandibular Joint Problems." *Physical Therapy: Basal Fact*. Vol. 4, no. 2, Jan. 1982.
21. Graber, T. M., D.M.D. "An Orthodontic Perspective After 75 Years." *American Journal of Orthodontics*. Vol. 6, no. 5, May 1976.
22. Greenberg, Martin S., D.D.S. "Mandibular Position and Upper Body Strength: A Controlled Clinical Trial." *Journal of American Dental Association*. Vol. 103, Oct. 1981.
23. Griffiths, Robert H., D.D.S. "Report of the President's Conference on the Examination, Diagnosis, and Management of Temporomandibular Disorders." *Journal of American Dental Association*. Vol. 106, Jan. 1983.
24. Hahn, George W., "Orthodontic Profiles—Edward H. Angle (1855–1930)." *American Journal of Orthodontics*. Vol. 51, no. 7, July 1965.
25. Howland, John, D.D.S., M.S. "The Winning Team: Orthodontics and Myofunctional Therapy." *International Journal of Oral Myology*. Vol. 5, no. 3, July 1979.
26. Kelsey, Charles C., D.D.S. "Milestones in the History of American Dentistry." *Pennsylvania Dental Journal*. Nov. 1976.

27. Lucia, Victor O., D.D.S., F.A.C.D. *Modern Gnathological Concepts*. St. Louis: C. V. Mosby Co., 1961.
28. Liddle, David W., D.D.S. "Second Molar Extraction in Orthodontic Treatment." *American Journal of Orthodontics*. Vol. 72, no. 76, Dec. 1977.
29. Luse, Robert R. "1776–1976 Bicentennial Progress of a Profession." *Journal of the New Jersey Dental Association*. Summer 1976.
30. Malvin, Ring E., D.D.S. "Paul Revere—Dentist and Our Country's Symbol of Freedom." *New York State Dental Journal*. Vol. 42, Dec. 1976.
31. Mathis, Bob, D.D.S., F.A.E.S. "Great Controversies in Dentistry." An article from the American Dental Association Library.
32. McCauley, Berton H., D.D.S. "American Dentistry in the Founding Years of the Republic." *Bulletin of the History of Dentistry*. Vol. 26, no. 2, Oct. 1976.
33. Mikhail, Mongi, B.D.S. and Harry Rosen, D.D.S. "History and Etiology of Myofacial Pain—Dysfunction Syndrome." *Journal of Prosthodontics Dentistry*. Vol. 44, no. 4, Oct. 1980.
34. O'Neill, Ynez V. "Speech and Speech Disorders in the Western Thought Before 1600." *Contributions to Medical History*. No. 3, Greenwood Press: Westport, CT, London, England, 1980.
35. "The Origin of Orthodontics." *Dental Survey*. Vol. 54, no. 9, Sept. 1978.
36. "Orthodontic Education (Pre–1900)." *American Journal of Orthodontics*. Vol. 52, no. 10, Oct. 1966.
37. Ottolengai, Rodriquez. "Orthodontic Profiles." *American Journal of Orthodontics*. Aug. 1958.
38. Paden, Elaine P. "ASHA in Retrospect: Fiftieth Anniversary Reflections." *ASHA Journal*. Vol. 17, no. 3, Mar. 1975.
39. ———. "ASHA in Retrospect: Fiftieth Anniversary Reflections." *ASHA Journal*. Vol. 17, no. 4, Apr. 1975.
40. ———. "ASHA in Retrospect: Fiftieth Anniversary Reflections." *ASHA Journal*. Vol. 17, no. 5, May 1975.
41. ———. "ASHA in Retrospect: Fiftieth Anniversary Reflections." *ASHA Journal*. Vol. 17, no. 6, June 1975.
42. ———. "ASHA in Retrospect: Fiftieth Anniversary Reflections." *ASHA Journal*. Vol. 17, no. 7, July 1975.
43. ———. "ASHA in Retrospect: Fiftieth Anniversary Reflections." *ASHA Journal*. Vol. 17, no. 8, Aug. 1975.
44. ———. "ASHA in Retrospect: Fiftieth Anniversary Reflections." *ASHA Journal*. Vol. 17, no. 9, Sept. 1975.

45. ———. "ASHA in Retrospect: Fiftieth Anniversary Reflections." *ASHA Journal*. Vol. 17, no. 10, Oct. 1975.
46. ———. "ASHA in Retrospect: Fiftieth Anniversary Reflections." *ASHA Journal*. Vol. 17, no. 11, Nov. 1975.
47. ———. "ASHA In Retrospect: Fiftieth Anniversary Reflections." *ASHA Journal*. Vol. 17, no. 12, Dec. 1975.
48. Paolino, Vincent J. "The Development of Dentistry as a Profession." *The New Dentist*. Vol. 2, Oct. 1980.
49. Proctor, James. "The Development of Orthodontics." *Dental Student's Magazines*. Apr. 1965.
50. Saffro, Dennis W. "Early Dentistry in the American Colonies." *Journal of Science,* California Dental Association. Vol. 36, Sept. 1968.
51. Schoolfield, Lucille D. "The Development of Speech Correction in America in the Nineteenth Century." *The Quarterly Journal of Speech*. Vol. 24, no. 1, Feb. 1938.
52. Schwarze, Claus W. "Expansion and Relapse in Long Follow-Up Studies." Orthodontic Department of University Dental Hospital in Cologne, 1979.
53. "75th Anniversary Issue." *The Dental Cosmos*.
54. Sproles, Alan C., D.D.S. "Dentistry In Colonial America." *Dental Survey*. Vol. 52, no. 4, Apr. 1976.
55. Strong, Robert H. W., D.D.S., M.D. "Highlights of Sixty-Four Years in Orthodontics." *Angle Orthodontics*. Vol. 44, no. 2, Apr. 1974.
56. ———. *Orthodontia*. 2nd ed., Philadelphia: Lea and Fabiger, 1969.
57. Thorum, Arden R., Ph.D. "Myofunctional Therapy and the Adolescent." *Impressions*. May 19, 1982.
58. "200 Years of Jewish Contribution to Dentistry in America." *Alpha Omega*. Vol. 69, no. 2, Sept. 1976.
59. Vego, Leroy, D.D.S., and Neal C. Murphy, D.D.S. "Early Orthodontic Diagnosis in General Dentistry: Method and Significance." *California Dental Association Journal*. Apr. 1982.
60. Weinberger, Bernhard W., D.D.S. *An Introduction to History of Dentistry*. Vol. 1, St. Louis: C. V. Mosby Co., 1948.
61. ———. "From 'Irregularities' of the Teeth to Orthodontics as a Specialty of Dentistry." *American Journal of Orthodontics*. Vol. 42, Mar. 1956.
62. ———. *Orthodontics*. St. Louis: C. V. Mosby Co., 1926.
Wilson, H. E., L.D.S. "Extractions of Second Permanent Molars in Orthodontic Treatment." *The Orthodontist*. Vol. 3, no. 1, Spring 1971.
63. Wise, Harry S. "Speech—The Overlaid Function." *The Quarterly Journal of Speech*. Feb. 1938.

Glossary

alveolar bone. The specialized bone structure that supports the teeth.

alveolar ridge. The bony ridge (alveolar process) of the maxilla or mandible that contains the alveoli (sockets of the teeth).

Bennett movement. The bodily lateral movement or lateral shift of the mandible resulting from the movements of the condyles along the lateral inclines of the mandibular fossae in lateral jaw movement.

bruxism. A clenching of the teeth associated with forceful lateral or protrusive jaw movements that results in rubbing, gritting, or grinding together of teeth, usually during sleep.

centric position. The position of the mandible in its most retruded (nonprotruded) relation to the maxilla.

centric relation. (1) The most retruded relation of the mandible to the maxilla when the condyles are in the most posterior unstrained position in the glenoid fossae from which lateral movement can be made at any given degree of jaw separation. (2) The most retruded relation of the mandible to the maxilla from which the patient can make lateral jaw movements. (3) The most posterior relation of the mandible to the maxilla at the established occlusal vertical dimension. (4) The most retruded unstrained relation of the mandible to the maxilla at a given degree of vertical opening. (5) The most retruded relation of the mandible to the maxilla at a given degree of vertical opening. (6) The most retruded voluntary relation of the mandible to the maxilla. (7) Any most retruded functional relation of the mandible to the maxilla.

condylar guidance; condylar guide. The mechanical device on an articulator that is intended to produce similar guidances in articulator movement as are produced by the paths of the condyles in the temporomandibular joints.

condyle. The rounded articular surface at the articular end of a bone.

crossbite. (1) dental: a dental crossbite exists when any tooth or group of teeth is captured by the lower arch. (2) mandibular: a mandibular crossbite exists any time the frenum of the maxilla is not in a straight line with the frenum of the mandible. (3) speech: (cosmetic shift) The jaw shifts into a crossbite position, laterally or anteriorly, during the production of certain speech sounds, usually the *s, z, sh, ch, zh,* and *j* sounds.

curve of Spee. Anatomic curvature of the occlusal alignment of teeth beginning at the tip of the lower cuspid and following the buccal cusps of the natural bicuspids and molars, continuing to the anterior border of the ramus as described by von Spee.

distal. (1) Away from the median sagittal plane of the face following the curvature of the dental arch. (2) In a dorsal direction from any given point of reference.

heredity. A predetermined, chromosome oriented blueprint for growth and development.

horizontal overlap. The projection of the upper anterior and/or posterior teeth beyond their antagonists in a horizontal direction.

jaw relation centric. (1) The jaw relation when the condyles are in the most posterior unstrained position in the glenoid fossae from which lateral movement can be made at any given degree of jaw separation. (2) The most posterior relation of the mandible to the maxilla at the established vertical dimension. (3) The relation of the mandible to the maxilla when the condyles are in their most posterior position in the glenoid fossae from which unstrained lateral movements can be made at the occluding vertical dimension normal for the individual.

mechanism. The entire oral structure plus the brain function.

mesial. (1) Toward or situated in the middle, as the median line of the body or the center of the dental arch. (2) Situated in the middle; median, toward the middle line of the body or toward the center line of the dental arch.

myofunctional therapy. A method for treating tongue thrust and other orofacial muscle (myo) dysfunction through exercises. Developed by a dentist and adopted as a therapy method by a small percentage of speech pathologists.

myosteodontics. A term coined by John Streicher in the 1980s to identify the discipline he developed for diagnosing and eliminating habits affecting muscle (myo), bone (osteo), and teeth (dontics).

occlusal. (1) Pertaining to the contacting surfaces of opposing occlusal units (teeth or occlusion rims). (2) Pertaining to the masticating surfaces of the posterior teeth.

occlusal equilibration. The modification of occlusal forms of teeth by grinding with the intent of equalizing occlusal stress, or producing simultaneous occlusal contacts, or of harmonizing cuspal relations.

occlusal plane, plane of occlusion. An imaginary surface that is related anatomically to the cranium and that theoretically touches the incisal edges of the incisors and the tips of the occluding surfaces of the posterior teeth. It is not a plane in the true sense of the word but represents the mean of the curvature of the surface.

occlusal position. The relationship of the mandible and maxilla when the jaw is closed and the teeth are in contact. This position may or may not coincide with centric occlusion.

oral. Pertaining to the mouth.

oral habit. A behavior that is persistent and consistent enough to produce a physical change in the oral structure.

oral structure. The oral cavity and contiguous structures.

orofacial. Pertaining to the oral and facial structures.

stress. A psychological or emotional pressure.

system. A functioning mechanism.

tension. Physical contraction of the muscle; muscle tightness.

tongue thrust. Any deviate tongue position that does not produce and maintain a normal occlusion.

tooth wear. Any wear not normal for the age of the patient.

Epilogue

This book is just the tip of the iceberg. It was written to announce my discovery and to give an overview of the concepts it entails. I have much more to share, so more books are forthcoming.

I am planning books to explain how my therapy helps in treating cases of cerebral palsy, cleft palate, hard-of-hearing, deafness, and stuttering. I intend to produce academic and scientific material for the dental and speech therapy professions. And I am planning a series of educational books for children and adults.

Sharing what I have learned during my forty-year career of treating habit and nonhabit conditions is my goal. More books will help, but my hope is that some university or organization will offer the funding and structure for a research-and-training center. We need to teach therapists from around the world because the problems I address are universal. All races, colors, and creeds are affected.

Everyone has an Aunt Hattie, so to speak—someone who influences one's choices. My Aunt Hattie started my quest to discover the causes of speech disorders. This book is her introduction.

To order additional copies of

THE PEBBLE *in the* Shoe

Call (520) 773-9795

or send $24.95 and a return address to

Streicher Publications
PO Box 765
Flagstaff, AZ 86002